Experiments in
# PHARMACEUTICAL
# CHEMISTRY

Second Edition

# Experiments in
# PHARMACEUTICAL CHEMISTRY

## Second Edition

Charles Dickson
Retired Instructor in Chemistry
Catawba Valley Community College
Hickory, North Carolina

CRC Press
Taylor & Francis Group
Boca Raton London New York

CRC Press is an imprint of the
Taylor & Francis Group, an **informa** business

CRC Press
Taylor & Francis Group
6000 Broken Sound Parkway NW, Suite 300
Boca Raton, FL 33487-2742

© 2014 by Taylor & Francis Group, LLC
CRC Press is an imprint of Taylor & Francis Group, an Informa business

No claim to original U.S. Government works

Printed on acid-free paper
Version Date: 20131227

International Standard Book Number-13: 978-1-4822-2508-2 (Paperback)

**Visit the Taylor & Francis Web site at**
**http://www.taylorandfrancis.com**

**and the CRC Press Web site at**
**http://www.crcpress.com**

# Contents

# Preface

The lack of instructional material for laboratory work in pharmaceutical chemistry has prompted the writing of this manual. While student interest and enrollment continues to increase, a search of the literature indicates little available material for laboratory work in this field.

This book is designed to meet those needs and, as such, will hopefully be considered by faculty in undergraduate and graduate schools, as well as by those in professional programs, particular medicine and pharmacy.

Twenty-two major topic areas are presented and supported by 75 experiments. Admittedly, this is far in excess of what could be covered in any one academic year, but the breadth and diversity of topics allows for a variety of choices.

Comments on the material are always welcome.

**Charles Dickson**
*Hickory, North Carolina*
*April 2013*

# The Author

**Dr. Charles Dickson** holds degrees from the University of Tampa, Wartburg Theological Seminary, Stetson University, and the University of Florida.

He has been a chemistry instructor in the nursing and prepharmacy program of Catawba Valley Community College in Hickory, North Carolina, for more than 20 years.

During his doctoral study at the University of Florida he held a National Institutes of Health Fellowship in pharmacology.

He is the author of three previous chemistry laboratory manuals, two for introductory courses and the third in medicinal chemistry. His articles in health care and alternative medicine have appeared in *Your Health* and *Mother Earth News* magazines, and he formerly wrote a syndicated column on these topics for the *New York Times*.

The author is now a retired chemistry professor.

# 1 Carbohydrates

## CARBOHYDRATES

The carbohydrates include substances that are constituents of the body's cells and of compounds that are important foods, plus products that have both medicinal and industrial uses. They are divided into these categories:

1. Monosaccharides—examples are glucose and fructose
2. Disaccharides, sometimes called oligosaccharides—examples of which are sucrose, lactose, and maltose
3. Polysaccharides—examples include starch, glycogen, and cellulose.

Medicinal products using carbohydrates include dextrose injection, calcium gluconate in both tablet and injection forms, heparin as an anticoagulent, and glycoside such as digitalis products used in the treatment of cardiac patients.

The experiments in this section are designed to illustrate chemical tests that identify the various types of carbohydrates, to perform extractions of sugars from various food products, and to isolate carbohydrate molecules from living tissues.

## EXPERIMENT 1: INTRODUCTION TO CARBOHYDRATES

Carbohydrates may be described as either polyhydroxy aldehydes or ketones or substances that yield these compounds upon hydrolysis; they are also known as aldoses and ketoses. The carbohydrate used by the body in energy metabolism is glucose—a hexose monosaccharide:

Straight chain                    Haworth

## REDUCING AND NONREDUCING SUGARS

Alkaline copper solutions readily oxidize many mono- and disaccharides. With the oxidation of the sugar there is a simultaneous reduction of the cupric ion ($Cu^2$) to a cuprous ion ($Cu^1$) compound, which can be identified easily by the color change from a deep blue to a red or reddish brown. The best-known solutions for these tests are

- Fehling's solution (an alkaline, $NaOH$, $CuSO_4$ solution with potassium sodium tartrate)
- Benedict's solution (an alkaline, $Na_2CO_3$–$CuSO_4$ solution in the presence of sodium citrate)
- Barfoed's solution (copper acetate in a .15 N acetic acid solution)

The following experiment will differentiate between reducing and nonreducing sugars. Place 10 mL of Benedict's solution in each of six test tubes. Add 50 mg of glucose to the first, fructose to the second, starch to the third, sucrose to the fourth, lactose to the fifth, and in the sixth tube starch with 10 mL of saliva. Shake well and allow the tubes to stand for 15 minutes. After this, place the six test tubes in a 600 mL beaker that is a third full of water and heat gently for 10 to 15 minutes. Record the results in the following table:

| Material | Color Reaction | Reducing (check) |
|---|---|---|
| 1. Glucose | | |
| 2. Fructose | | |
| 3. Starch | | |
| 4. Sucrose | | |
| 5. Lactose | | |
| 6. Starch/saliva | | |

Explain how the structures of certain sugars enable them to reduce the copper ion.

Explain the reaction of the starch/saliva tube in comparison to the tube with starch only.

## BARFOED'S QUALITATIVE TEST

Place 10 mL of Barfoed's solution in each of four test tubes. Place 50 mg of glucose, fructose, maltose, and sucrose, in each of these tubes. Warm gently in a water bath and carefully time each reaction. Explain the results.

## EXPERIMENT 2: COLOR TESTS OF CARBOHYDRATES

Because of their variant molecular structures the carbohydrates form numerous color complexes with specific reagents, thus helping the analyst to identify them, sometimes not specifically, but at least as to class, such as, for example, mono- or disaccharide and hexose or pentose.

## MOLISCH TEST

In this test concentrated $H_2SO_4$ forms furfurals due to its dehydrating action on the carbohydrates, and the furfurals or intermediate products between the carbohydrates and the furfurals condense with α-naphthol to form colored products. α-Naphthol's structure is

A negative result by this reaction is very good evidence of the absence of carbohydrates, but a positive test is simply an indicator of the probable presence of carbohydrates. Among the substances that give the test are various furfurals, glycuronic acid, glycoaldehyde, erythrose, and glyceraldehyde. The reaction can be made more valuable by examining the absorption spectra of the alcoholic solutions of the colored condensation products.

In each of four test tubes place 1 mL of .02 M glucose, .01 M xylose, 0.7% starch, and .01 M furfural. Add 2 drops of fresh 5% α-naphthol solution in alcohol directly to each substance. Mix each well, then allow 3 mL of concentrated $H_2SO_4$ to flow down the side of each tube to form a lower layer of acid.

Results:

Glucose _____
Xylose _____
Starch _____
Furfural _____

## SELIWANOFF TEST

This is an indicative test for hexoses since keto-hexoses, keto-trioses, keto-tetroses, d-oxygluconic acid, and formose also react positively. The color developed here is due to the formation of 4-hydroxy-methyl-furfural, and the reaction of this with resorcinol in the presence of a 12% HCl solution. Resorcinol's structure is

Prepare fresh Seliwanoff by adding 3.5 mL of 0.5% fresh resorcinol to 12 mL of concentrated HCl and dilute to 35 mL with distilled water. Mix well and measure off 5 mL portions into test tubes containing, respectively, 1 mL of 0.1 M glucose, 0.1 M galactose, 0.1 M levulose, 0.1 M xylose, and 0.1 M furfural. Mix well and immerse the tubes in boiling water. Observe the color changes every 5 minutes during 15 minutes of boiling.

Results:

Glucose _____
Galactose _____
Levulose _____
Xylose _____
Furfural _____

## TOLLEN'S PHLOROGLUCINOL TEST

This reaction is positive for pentoses but not a specific indication of them since hexoses and glycuronic acid also react. The distinction between the groups is in a carefully conducted test. If the cherry red color develops very promptly, and if this is soluble in amyl alcohol and the alcoholic solution shows a specific absorbance band consistent with resource data, then it can be considered a fairly reliable test for pentoses and glycuronic acid. The colored compounds formed are condensation products of the various intermediate and final acid decomposition products from carbohydrates with the phenolic compound phloroglucinol, which is a 1.3.5 trihydroxy benzene:

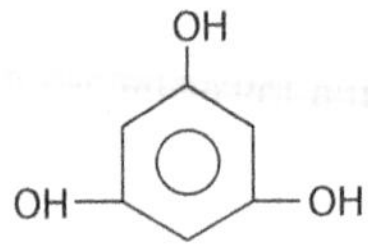

Prepare a fresh reagent by mixing 10 mL of concentrated HCl with 2 mL of the phloroglucinol solution and dilute to 18 mL with water. Measure off 5 mL portions of this solution and add, respectively, to tubes containing 0.5 mL of .02 M xylose, 0.5 mL of .02 M levulose, 0.5 mL of .02 M glucose, and 0.5 mL of .02 M furfural solutions. Place all but the last tube in the boiling water and note the color changes over 15 minutes.
  Result:

Xylose _____
Levulose _____
Glucose _____
Furfural _____

## TOLLEN'S ORCINOL TEST

Much of the information derived from the phloroglucinol test also holds true for the orcinol reaction with similar differences being observed, but the latter now yield greenish-blue colors instead of cherry red. Orcinol is a 3,5 dihydroxy toluene:

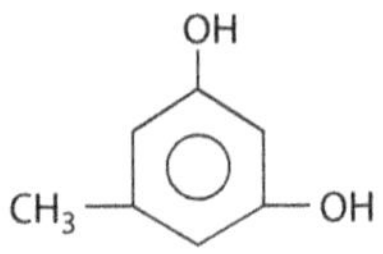

Prepare a fresh reagent by mixing 10 mL of concentrated HCl with 2 mL of a 6% aqueous solution of orcinol and dilute it to 18 mL with water. Now measure off 5 mL portions of the orcinol reagent into test tubes that contain, respectively, 0.5 mL of .02 M xylose, .02 M levulose, .02 M glucose, and .02 M furfural solutions. Place all but the last tube in boiling water and observe the color changes. Examine the absorption spectra.

Results:

Xylose      _____

Levulose _____

Glucose   _____

Furfural  _____

## EXPERIMENT 3: PRODUCING ETHANOL FROM SUCROSE

One of the most important commercial reactions of carbohydrates is the fermentation of certain sugars to form ethyl alcohol. We will use sucrose, a disaccharide with the formula $C_{12}H_{22}O_{11}$. It has one glucose molecule combined with fructose. The enzyme zymase converts the sugar to alcohol and carbon dioxide. Prior to starting the experiment you will need to make up a quantity of Pasteur's salts. This consists of 2.0 g of potassium phosphate, 0.20 g of magnesium sulfate, 0.20 g of calcium phosphate, and 10.0 g of ammonium tartrate dissolved in 860 mL water.

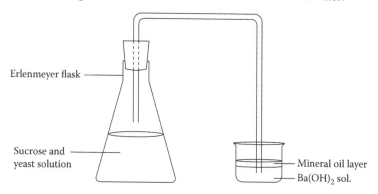

Erlenmeyer flask

Sucrose and yeast solution

Mineral oil layer

$Ba(OH)_2$ sol.

Place 40 g of sucrose in a 500 mL Erlenmeyer flask. Add 350 mL of water, 35 mL of Pasteur's salts, and 15 g of baker's yeast. Shake vigorously and fit the flask with a one-hole rubber stopper with a glass tube leading to a beaker containing a solution of barium hydroxide. Protect the barium hydroxide from air by adding some mineral oil to form a layer over it. A precipitate of barium carbonate will eventually form indicating that $CO_2$ is being evolved:

$$Ba(OH)_2 + CO_2 \rightarrow BaCO_3 + H_2O$$

Allow the mixture to stand for 1 week before proceeding. After this period of time, carefully remove the rubber stopper and siphon off the liquid from the flask, being careful not to disturb the sediment. Add 46 g of anhydrous potassium carbonate for each 100 mL of solution, which, in effect, saturates the solution. Transfer to a distillation apparatus and slowly heat until a temperature of 78°C is reached. At this point discard any distillate collected and begin to collect the ethanol. Do not heat beyond 88°C. Discard the residue in the flask and measure the volume of ethanol collected.

Calculate the percentage yield of ethanol based on an average product yield of 85% water and 15% water.

Remember, when a disaccharide is used in fermentation, the enzyme in yeast first converts into a mixture of invert sugar (glucose and fructose), which then undergoes fermentation according to the equation:

$$C_6H_{12}O_6 \rightarrow 2\ C_2H_5OH + 2\ CO_2$$

$$\frac{\text{Actual yield}}{\text{Theoretical yield}} \times 100 = \underline{\hspace{1cm}} \%\ \text{yield}$$

## EXPERIMENT 4: THE ISOLATION OF LACTOSE FROM MILK

Lactose, which has the popular name of "milk sugar," can be isolated from milk by a fairly simple laboratory procedure. It is a disaccharide with the structural formula of:

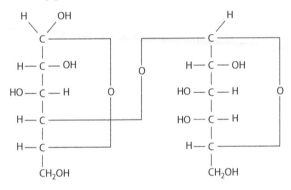

Place 200 mL of skimmed (nonfat) milk in a 600 mL beaker and warm it to 40°C, then add dropwise some 10% acetic acid until there is a significant precipitate of casein. Do not add excess acid as it may hydrolyze the disaccharide lactose to the monosaccharides glucose and galactose. Stir the casein until it forms a large amorphous mass and remove it from the mixture. You may want to save it for another experiment involving proteins. The clear solution now contains the lactose you are seeking to isolate. Add 5 g CaCO$_3$ to the clear solution.

Heat this mixture to a gentle boil for about 10 minutes. This should result in the nearly complete precipitation of the albumins. Filter the hot mixture by vacuum to remove the precipitated albumins plus any remaining CaCO$_3$ and concentrate the filtrate in a 600 mL beaker, using a water bath, to about 30 mL. Do not boil too vigorously or excessive foaming and boiling over may occur. You may also want to place marble chips in the mixture to prevent bumping. Now add 175 mL of 95% ethanol and 2 g of decolorizing charcoal to the hot solution. After it has been mixed well, filter the warm solution by vacuum through a layer of wet filter aid (using diatomaceous earth). Make sure the filtrate is clear.

Transfer the filtrate to a flask and allow it to stand for several days. This provides time for the lactose to crystallize on the walls and bottom of the flask. Dislodge the crystals and vacuum filter them. Wash the product with 10 mL of 25% ethanol. Weigh the product after it is thoroughly dry.

Result: _____ g.

Using the density of skimmed milk as 1.03 g/mL, calculate the percentage yield of lactose and record the calculations below.

## BENEDICT'S TEST ON LACTOSE

To 1 mL solutions of 1% lactose, glucose, galactose, and distilled water add 5 mL Benedict's solution and place all four tubes in a boiling water bath for 3 minutes.

Record results:

Lactose            _____
Glucose            _____
Galactose          _____
Distilled $H_2O$ _____

# EXPERIMENT 5: OBTAINING MUCIC ACID FROM GALACTOSE

When a strong $HNO_3$ solution is heated with an aldose hexose, it oxidizes to the corresponding dicarboxylic acid. Galactose, for example, yields mucic acid, which is sparingly soluble in water:

$$C_6H_{12}O_6 + 6\ HNO_3 \rightarrow COOH(CHOH)_4\ COOH + 4\ H_2O + 6\ NO_2$$

To 25 mL of 0.1 M lactose solution in a 100 mL beaker, add 10 mL of distilled water and 15 mL $HNO_3$. Mix well and heat on a steam bath until concentrated to about one-third the original volume. Then add 10 mL of distilled water and mix well. Set aside in a cool place until the next day. Filter the mucic acid crystals, dry, and weigh the product.

Result: _____ g of mucic acid.

Calculate the percentage yield by comparing the actual yield to the theoretical yield based on the formula stoichiometry.

$$\frac{\text{Actual yield}}{\text{Theoretical yield}} \times 100 = \text{_____ \% yield}$$

a persistent faint pink color. Each mL of the 0.5 N NaOH is the saponification titer and is equivalent to 15.52 mg of methoxy ($-OCH_3$) on an undried basis.

Result: _____ mL (NaOH) × 15.52 = _____ mg ($-OCH_3$)

## ASSAY FOR GALACTURONIC ACID

Each mL of the 0.5 N NaOH used in the total titration (initial plus saponification) is the equivalent of 97.07 mg of galacturonic acid ($C_5H_9O_5COOH$) on an undried basis.

Results:

| | | |
|---|---|---|
| Initial titer | _____ | mL NaOH |
| Saponification titer | _____ | mL NaOH |
| Total | = _____ | mL NaOH |
| × 97.07 | = _____ | mg Galacturonic acid |

Consult the literature such as *Remington's Practice of Pharmacy,* the USP/NF, and pharmacognosy texts, and list some of the commercial and pharmaceutical uses for pectin.

## PREPARATION OF AMYLOSE AND AMYLOPECTIN FROM POTATO STARCH

Many different methods have been used to fractionate starch into its branched and linear components. The method most suitable for one kind of starch is not necessarily the best for another. In the following experiment, potato starch is separated into its principal components—amylose and amylopectin—by the addition of dilute alkali followed by neutralization. Amylose remains in solution while amylopectin forms a gel. The polysaccharide starch molecule may be represented as a chain of glucose molecules:

*Preparation of amylose.* Place 250 g of previously sliced and peeled potatoes into a beaker with 200 mL of 1% NaCl solution. Filter the slurry through fine muslin and discard the liquid. Return the slurry to another 200 mL of 1% NaCl solution and repeat the filtration process. Wash the starch with 100 mL of 0.1 M NaOH, and then wash three times with distilled water.

Next add 20 g of the drained starch to 800 mL of a 0.157 M NaOH solution to which 70 mL of water have been added. The mixture should be stirred gently until it clears. The emphasis here is on gentle stirring so as not to break the pectin gel. This alkaline solution should then stand for 5 minutes after which 250 mL of a 5% NaCl solution are added and the dispersion neutralized with sufficient 1.0 M HCl to create a solution of pH 6.5–7.5, which is monitored by an external indicator. Allow this to stand overnight. The gel will settle, occupying about one-third of the volume. The amylose solution forms the supernatant, which can then be siphoned off.

The amylose solution should not be stored more than a few days. Add 40 mL of n-butanol in order to precipitate the amylose. Centrifuge in order to concentrate the precipitate. Siphon off the supernatant and vacuum filter the amylose.

Record results:

Weight of amylose _____ g.
Percentage of amylose in potato starch _____ %

*Purification of amylopectin.* The amylopectin gel should be centrifuged at 8000 rpm for 30 minutes. Discard the resulting supernatant and add 100 mL of a 1 of a 1% NaCl. Allow the mixture to stand overnight at 18°C and centrifuge at 8000 rpm as before.

Weigh the gel and record results:

Weight of amylopectin _____ g.

The gel can be stored in water since it is insoluble and can also be freeze-dried and stored.

## EXPERIMENT 6: PECTINS

Pectins are polygalacturonic acid esters and can have molecular weights from 20,000 to 400,000. These polysaccharides are present in many fruits, particularly in orange and lemon rinds, where they may constitute as much as 30% of the total weight. The galacturonic acid unit has a structure of:

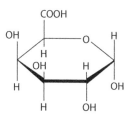

Grind 20 g of orange rind (no outer peel) and place in a 250 mL flask attached to a Soxhlet extractor. Add 100 mL of dilute HCl and boil gently for an hour. This process dissolves the pectin from the rind and puts it in solution. After cooling, add 100 mL of 95% ethanol, which precipitates the pectin. Filter, dry, and weigh. The powder is yellowish-white in color, almost odorless, possessing a mucilaginous taste.

Result: _____ g.

Calculate the pectin% in the rind.

Result: _____ %.

### ASSAY OF PECTINS FOR METHOXYL GROUPS

Transfer 5 g of the pectin to a 500 mL beaker and agitate with a magnetic stirrer for 10 minutes with 5 mL of dilute HCl and 100 mL of ethanol (60%). Transfer to a fritted glass filter tube and wash with six 15 mL portions of HCl-60% ethanol to free the mixture of chlorides. Dry for 1 hour in a drying oven at 105°C, cool, and weigh. Now transfer exactly one-tenth of the total net weight of the dried sample to a 250 mL Erlenmeyer flask and moisten the sample with 2 mL alcohol. Add 100 mL of recently boiled and cooled $H_2O$ stopper, and swirl occasionally until the pectin is completely dissolved. Add 5 drops phenolphthalein t.s. and titrate with 0.5 N NaOH, and record the results as the initial titer. Add 20 mL of 0.5 N NaOH again, stopper, and swirl. Let stand for 15 minutes. Add exactly 20 mL of 0.5 N HCl and shake until the pink color disappears. After adding 3 drops of phenolphthalein t.s. titrate with 0.5 N NaOH to a persistent faint pink color. Each mL of the 0.5 N NaOH is the saponification titer and is equivalent to 15.52 mg of methoxy ($-OCH_3$) on an undried basis.

Result: _____ mL (NaOH) × 15.52 = _____ mg ($-OCH_3$)

## ASSAY FOR GALACTURONIC ACID

Each mL of the 0.5 N NaOH used in the total titration (initial plus saponification) is the equivalent of 97.07 mg of galacturonic acid ($C_5H_9O_5COOH$) on an undried basis.
    Results:

| | | |
|---|---|---|
| Initial titer | _____ | mL NaOH |
| Saponification titer | _____ | mL NaOH |
| Total | = _____ | mL NaOH |
| × 97.07 | = _____ | mg galacturonic acid |

Consult the literature such as *Remington's Practice of Pharmacy*, the USP/NF, and pharmacognosy texts, and list some of the commercial and pharmaceutical uses for pectin.

# EXPERIMENT 7: SEPARATION OF SUGARS BY COLUMN CHROMATOGRAPHY

Charcoal has a long history of usage associated with laboratory and industrial processes for the purification of sugar. Chemists in the 1930s began to perform experiments to separate such sugars as glucose, maltose, and raffinose from each other.

Make up 10% solutions of glucose, maltose, and raffinose by placing 1 g of each sugar in 9 g (mL) of water and shake well until all the sugar is dissolved. Mix the three solutions and pour down a chromatography column measuring approximately 230 mm in length and 34 mm in diameter, which has been packed by an adsorbent mixture of equal parts by weight of DARCO G-60 and Celite (special forms of charcoal and diatomaceous earth).

The sugar solution is then adsorbed on the column by successive introduction of 800 mL water, 1.5 liters of 5% ethanol, and 700 mL of 15% ethanol. The effluent is collected in 100 mL fractions, and each collection is completed prior to the introduction of the new developer.

Solutions are then crystallized by boiling off the water and the specific sugar identified polarimetrically and by specific melting point.

Record results:

1st sugar in effluent _____
2nd sugar in effluent _____
3rd sugar in effluent _____

## EXPERIMENT 8: ISOLATION OF CHITIN AND GLUCOSAMINE

Chitin is a polyglucosamine widely distributed in vertebrates. Because of their ready availability and low protein content, decalcified shells are the most suitable source of chitin. Crab and lobster shells are commonly employed.

The structural formula for Chitin $(O_8H_{13}O_5)_n$ is:

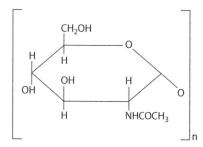

*Procedure for isolating chitin.* Obtain some crab shells, clean, and dry them in an oven at 100°C. Break them into small pieces and place 100 g of these fragments into 1.5 liter of ".0 M HCl at room temperature overnight. The resulting rubbery residues are then thoroughly washed with water and dried at 100° C. Weigh the dried product and calculate the percentage of chitin in the crab shells

$$\frac{\text{Weight of chitin residue} \cdot \text{g}}{\text{Weight of crab shells} \cdot 100 \cdot \text{g}} \times 100 = \underline{\quad\quad} \%$$

*Procedure for isolating glucosamine.* Glucosamine has a molecular weight of 179.17 and finds use in pharmacology as an antiarthritic drug. Its molecular structure is

Place 400 mL of concentrated HCl with the dried chitin and heat on a water bath for 2 and 1/2 hours. Separate the solution from the black sludge by filtering through a bed of celite. Stir the brown filtrate for 30 minutes at 60°C with activated charcoal and filter again through the celite. The filtrate should now have a pale yellow color. Concentrate the solution to a volume of 75 mL at 50–60°C under reduced temperature. Add 300 mL of ethanol to this concentration and store for 24 hours at room temperature.

The next day remove the product by filtration and wash successively with ethanol and ether and air dry. Record your yield of glucosamine _____ g.

Further purification may be affected by dissolving the product in a minimum volume of hot water and adding 80 mL of ethanol. D-glucosamine HCl, crystallized as the anomer, may then be recovered by precipitation with ether.

Check with local pharmacies and list some of the manufacturers of glucosamine, including those who combine it with chondroitin and MSM as antiarthritic agents.

## EXPERIMENT 9: ISOLATION OF GLYCOGEN FROM LIVER TISSUE

Glycogen, or animal starch, is stored in the liver tissue and can be measured by a characteristic reaction with iodine. Glycogen is composed of branched chains of glucose units. Ultimately, in metabolism, it is hydrolyzed by enzymes to glucose. When glycogen is needed for energy, it is first converted to pyruvic acid according to the formula:

$$(C_6H_{11}O_5)_x \longrightarrow 2_x \; CH_3 \overset{O}{\overset{\|}{-C}} \overset{O}{\overset{\|}{-COH}}$$

To isolate glycogen from liver tissue, grind 10 g of the liver into small pieces and mix with 25 g of a 60% KOH solution. Agitate on a magnetic stirrer for 45 minutes. Add 20 mL water, boil for 10 minutes, and filter. Mix 2 mL of the filtrate with 0.5 g of KI, 7 mL of ethanol, and a few drops of phenolphthalein as an indicator. Add 50% HCL, drop by drop, until the indicator is decolorized.

Centrifuge to throw down the precipitate and discard the supernatant. Dry the powder and weigh.

Weight of glycogen _____ g.

To a few milligram of the glycogen obtained add a drop of 5% $I_2$ in 10% KI solution. Read at 470 μm against a reagent background.

## SELECTED REFERENCES—CARBOHYDRATES

Aspinwall, G. O. *The Polysaccharides*. Academic Press, 1983.

Binkley, W. *Modern Carbohydrate Chemistry*. Dekker, 1987.

Bonner, J. *Plant Biochemistry*. Academic Press, 1950.

Garg, H. *Carbohydrates: Chemical, Biological, and Medical Applications*. Elsevier, 2008.

Ortega, M. *New Developments in Medicinal Chemistry*. Nova Science, 2009.

Guthrie, R. D. *Introduction to the Chemistry of Carbohydrates*. Clarendon, 1974.

Lindhurst, T. *Essentials of Carbohydrate Chemistry and Biochemistry*. Wiley, 2007.

Manners, D. J. *Biochemistry of Carbohydrates*. Johns Hopkins, 1978.

Morrison, R. *Organic Chemistry*. Allyn & Bacon, 1987.

Pigman, W. *Carbohydrates*. Wiley, 1935.

Roehrig, K. L. *Carbohydrate Chemistry*. AVI Books, 1984.

Sinnott, M. *Carbohydrate Chemistry and Bioohemistry*. RSC Publishing, 2013.

Smith, E. L. *Principles of Biochemistry*. McGraw-Hill, 1983.

Snell, F. *Colorometric Mehtods of Analysis*. Van Nostrand, 1961.

Stick, Robert *Carbohydrates, Essential Molecules*. Elsevier, 2012.

Tipson, S. *Textbook of Biochemistry*. Wiley, 1935.

Whistler, R. L. *Methods in Carbohydrate Chemistry*. Academic, 1980.

Witzak, L. *Carbohydrate Synthons in Natural Products Chemistry*. A.C.S., 2003.

# 2 Lipids

## LIPIDS

Lipids include not only true fats but also substances that are chemically related to fats or related to them because of common solubilities as well as possible biological relationships. There are five categories of lipids:

1. Fats—esters of fatty acids and glycerol
2. Waxes—esters of fatty acids and alcohols
3. Phospholipids—fats that contain, in addition, phosphoric acid and nitrogenous groups
4. Cerebrosides—combinations of fatty acids, sugars, and nitrogenous substances
5. Sterols—hydrogenated phenanthrene derivatives

The lipids are generally soluble in ether and allied solvents, while carbohydrates are practically insoluble.

Some examples of lipids used as medicines include lipid-soluble vitamins, A, D, E, and K. Also the phospholipid, lecithin, and commercial glycerol used as a sweetening agent and skin emollient, plus the long list of steroid drugs.

Experiments in this section involve the hydrolysis of fats, the calculation of saponification and iodine numbers of oils, steroid experiments, and the preparation of a narcotic drug from botanical sources.

## EXPERIMENT 10: THE ALKALINE HYDROLYSIS OF FAT

Lipids include not only true fats but also substances chemically related to fats (like lecithin) or related to them because of common solubilities and possible biological relationships (like cholesterol). True fats and fixed oils are esters of glycerol and the higher fatty acids. The classification of lipids include

1. Fats—esters of fatty acids and glycerol
2. Waxes—esters of fatty acids and certain alcohols
3. Phospholipids—including phosphoric acid and nitrogen groups
4. Cerebrosides—fatty acids, sugars, and nitrogenous groups
5. Sterols—hydrogenated phenanthrenes like ergosterol and cholesterol

The lipids are soluble generally in ether and other solvents, whereas carbohydrates and proteins are practically insoluble in such solvents.

## Separation of Fatty Acids and Glycerol

Weigh out 10 g of rendered fat into a 500 mL flask and add 100 mL of 95% ethyl alcohol and 10 mL of a 40% NaOH solution. Fit with a water-cooled reflux condenser and boil for an hour on a steam bath. Transfer to a porcelain dish and evaporate to about 25 mL in a steam bath in order to remove the alcohol. Next, dissolve completely in 300 mL of hot water. To separate the fatty acid, acidify the hot aqueous solution with HCl. Heat, and when a sharp separation of an upper layer of fatty acids occurs, siphon the aqueous layer and proceed to evaporate to dryness for glycerol tests. Transfer the hot liquid fatty acid layer to a separatory funnel and shake four times with 150 mL portions of hot water. This removes the excess inorganic salts, glycerol, and HCl from the solution. Finally, pour the molten fatty acid into a small beaker and allow to clear in the molten condition. After a clear fatty acid layer is obtained, carefully measure 1 mL with a dry pipette, then transfer to a 150 mL conical flask containing 50 mL of 95% alcohol previously neutralized (with phenolphthalein) by 0.1 N NaOH. Warm, if necessary, to dissolve the fatty acids and titrate. Assuming that 1 mL of the molten fatty acids weighs 0.85 g, calculate the molecular weight of the fatty acid mixture on the basis of your titration values.

$$C_3H_5(OOC_{18}H_{35})_3 \ 3 \ NaOH \ C_3H_5(OH)_3 \ 3 \ NaOOC_{18}H_{35}$$

Result of calculations: _____.

## Extraction of Glycerol

Evaporate the first aqueous phase from the beginning of the previous experiment to dryness. Moisten the entire mass again with ethyl alcohol and evaporate again on a steam bath. Now extract the dry residue, consisting mainly of NaCl and glycerol with three 35 mL portions of alcohol, warming it each time and filtering if necessary. Combine these alcohol filtrates and evaporate in a small dish on the water bath to a syrup. Measure the amount of glycerol.

     Result: _____ mL.

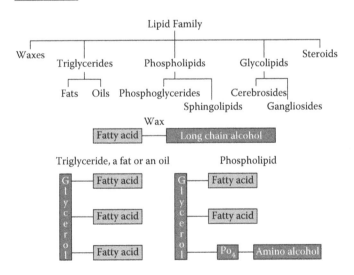

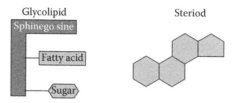

Glycolipid                                    Steriod

## EXPERIMENT 11: DETERMINING THE SAPONIFICATION NUMBER OF AN OIL

The saponification number of an oil or fat represents the number of milligram of KOH needed to saponify 1 g of the substance. Roughly speaking, this varies with the molecular weight of the fat or oil. Some average accepted saponification values for common oils include:

| | |
|---|---|
| Olive oil | 185–196 |
| Linseed oil | 192–195 |
| Cottonseed oil | 193–195 |
| Sesame oil | 188–193 |
| Coconut oil | 246–260 |
| Palm oil | 242–250 |
| Peanut oil | 188–195 |
| Corn oil | 187–196 |

Place 2 g of one of the oils above in a 250 mL flask and add 25 mL of a 0.5 N alcoholic KOH solution. Insert into the neck of the flask by means of a perforated stopper an air condenser and heat the flask over a water bath for 30 minutes. Then add 1 mL of phenolphthalein t.s. and titrate the excess of KOH with 0.5 N HCl. Now run a blank test, using exactly the same amount of 0.5 N alcoholic KOH. The difference between the number of milliliter of 0.5 N HCl consumed in the actual test and in the blank test multiplied by 28.05 is the saponification value of the oil being tested. Run two more oil samples through this procedure and record results.

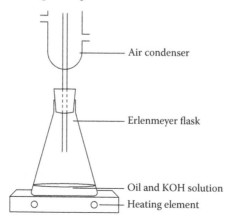

1. mL HCl used in first oil test—mL HCl in blank × 28.05 saponification number of first oil _____
2. mL HCl used in second oil test—mL HCl in blank × 28.05 saponification number of second oil _____
3. mL HCl used in third oil test—mL HCl in blank × 28.05 saponification number of third oil _____

Finally, compare your values of each specific oil sample with the average value presented in the literature.
Result:

First oil sample    _____ to _____
Second oil sample   _____ to _____
Third oil sample    _____ to _____

## EXPERIMENT 12: DETERMINING THE IODINE NUMBER OF AN OIL

The iodine value of a fat or oil represents the number of grams of iodine absorbed under the prescribed conditions by 100 g of the oil or fat. For these tests, we will use the Hanus method. The iodine number of some common oils are as follows:

| | |
|---|---|
| Olive oil | 79–88 |
| Linseed oil | 173–201 |
| Cottonseed oil | 108–110 |
| Coconut oil | 8–10 |
| Palm oil | 13–17 |
| Sesame oil | 103–122 |
| Corn oil | 109–133 |
| Peanut oil | 84–104 |
| Soybean oil | 127–138 |
| Safflower oil | 140–150 |

Introduce 250 mg of an oil sample into a glass-stoppered flask or bottle of 250 mL capacity and dissolve it in 10 mL chloroform and 25 mL of iodobromide t.s. Stopper the vessel securely and allow it to stand 30 minutes protected from light. Add 30 mL of KI t.s. and 100 mL water and titrate the liberated iodine with 0.1 N $Na_2S_2O_3$ solution, shaking thoroughly after each addition of the thiosulfate or agitating on a magnetic stirrer. When the iodine color becomes pale, add 1 mL of starch t.s. and continue to titrate with 0.1 N $Na_2S_2O_3$ until the blue color is discharged. Run a blank containing no oil through the same procedure and titrate in the same manner.

The difference between the number of millilitre of 0.1 N $Na_2S_2O_3$ consumed by the blank test and the actual test multiplied by 1.269 and divided by the weight in grams of the sample tested is the iodine number of the oil sample. Repeat this procedure twice again with two other oil samples and record data below.

1. mL $Na_2S_2O_3$ used in first oil test – mL used in blank test × 1.269 ÷ weight(g) of sample _____
2. mL $Na_2S_2O_3$ used in second oil test – mL used in blank test × 1.269 ÷ weight(g) of sample _____
3. mL $Na_2S_2O_3$ used in third oil lest – mL used in blank test × 1.269 ÷ weight(g) of sample _____

Finally, compare your values of each specific oil sample with the average value presented in the literature.

Result:

First oil sample      _____ to _____
Second oil sample    _____ to _____
Third oil sample      _____ to _____

## EXPERIMENT 13: ISOLATION OF CHOLESTEROL FROM GALLSTONES

Cholesterol is an example of the steroid class of lipids characterized by a cyclopentanoperhydrophenanthrene ring system. It occurs in brain tissue, gallstones, and blood. Its structure is

To begin the isolation process weigh 8 g of pulverized gallstones and place them in a 125 mL flask. Add 40 mL of diethyl ether and heat the mixture on a steam bath until the cholesterol is dissolved. Filter the brownish yellow solution through a fluted funnel while it is still hot and add a little ether to replace that lost through evaporation. The brown residue that collects on the filter paper is bilirubin, a bile pigment derived from hemoglobin with the structure:

Dilute this filtrate with 40 mL methanol, add a little decolorizing agent, and heat the mixture on a steam bath. Preheat a funnel, then filter the hot solution through a fluted filter paper. Reheat the greenish yellow filtrate and add just enough water, dropwise, to make the solution cloudy. The solution is now saturated at the boiling point and cholesterol will crystallize upon cooling. Collect the crystals by vacuum filtration using a small Buchner funnel as shown in the diagram.

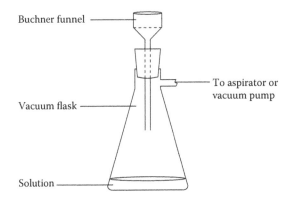

Wash the crystals with cold methanol and allow them to stand for a time in the open Buchner funnel to allow the solvent to evaporate. Weigh the dried cholesterol crystals.

     Result: _____ g. (Hint: gallstones are about 60% cholesterol.)

Determine the melting point of the crude, impure cholesterol. Record.

     Result: _____ °C.

Check results with the Merck Index. Save the remainder for the next experiment.

**Note:** Additional procedures would be needed to achieve isolation of absolutely pure cholesterol so your melting point determination will not match the literature exactly.

## EXPERIMENT 14: COLOR REACTIONS OF CHOLESTEROL

A number of time-honored tests can be run with your cholesterol sample from the previous experiment, which include the iodine-sulfuric acid test, the Salkowski reaction, the Liebermann–Burchard reaction, and the Rosenheim trichloroacetic acid test.

### IODINE–SULFURIC ACID

Prepare a mixture of 10 mL concentrated $H_2SO_4$ and 2 mL $H_2O$. Mix well and allow to cool. Evaporate to dryness in two test tubes 1 mL of 0.1% cholesterol and 1 mL of 0.01% ergosterol solutions in chloroform. After the chloroform is removed completely and the tubes have cooled, add 5 mL of the prepared $H_2SO_4$ solution 10 each tube.

Record results _____ and _____.

Now add 1 drop of 0.1 N I in KI and record results of each tube.
   Results: _____ and _____.

Add 2 drops more of the I/KI reagent.
   Results: _____.

## SALKOWSKI REACTION

Into two absolutely dry test tubes, transfer 1 mL of a 0.1% cholesterol solution and a 0.1% ergosterol solution in chloroform. Add 1 mL of concentrated $H_2SO_4$ to each.

   Results:
   With cholesterol _____
   With ergosterol _____

**Note:** Cholesterol and its related steroid, ergosterol, will give very different color reactions in the above two tests.

## LIEBERMANN–BURCHARD REACTION

This reaction is used for the quantitative estimation of cholesterol. To two absolutely dry test tubes, transfer 1 mL of a 0.1% solution of cholesterol and 1 mL of a 0.01% ergosterol solution in chloroform. Add 5 drops of acetic anhydride to each tube. Mix each and add 1 drop of concentrated $H_2SO_4$ to each tube. Note the color changes and intensities in each tube for the first 5 minutes.

   Results:
   Cholesterol _____
   Ergosterol _____

Set aside overnight in a dark place and record results next day.

   Results:
   Cholesterol _____
   Ergosterol _____

As an added project you may want to plot concentrations using a spectrophotometer.

## ROSENHEIM TRICHLOROACETIC ACID

To 1 mL of the 0.1% cholesterol and 0.1% ergosterol solutions in chloroform, add 1 mL of trichloroacetic acid solution (9:1 mix of TCA:$H_2O$) and record results.

   Result:
   Cholesterol _____
   Ergosterol _____

Record absorption band readings of each solution at the time of the tests and redo these readings the next morning. Record the results:

## EXPERIMENT 15: ISOLATION OF TRIMYRISTIN FROM NUTMEG

The nutmeg, known botanically as *Myristica fragrans,* is the vegetable oil source of trimyristin, so it is in the lipids section; its isolation is described here using crushed nutmeg powder as the source.

The formula for trimyristin is

$C_{45}R_{86}O_6$ known also as glyceryl trimyristate

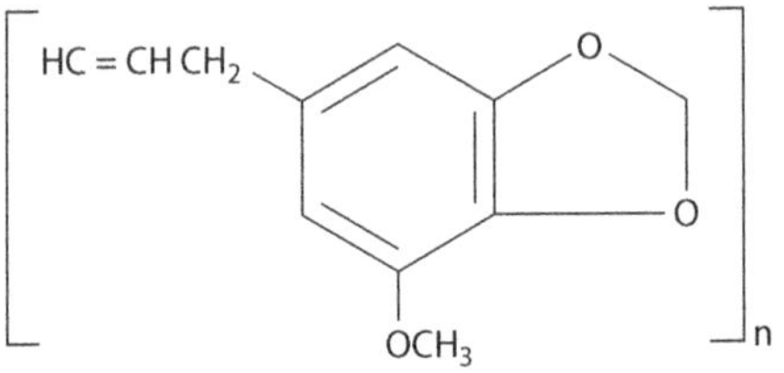

The compound has a molecular weight of 723.14 with a melting point of 56° C. It is insoluble in water, as are other lipids, but soluble in alcohols, benzene, chloroform, and ether. It has narcotic activity.

Preparation: Place 30 g of crushed nutmeg and 200 mL of chloroform in a reflux apparatus and heat for 90 minutes in a water bath. Next filter the nutmeg/chloroform solution and distill the solvent, which will leave a semisolid residue. This residue is then dissolved in 200 mL of 95% ethanol. Upon cooling, crystals of trimyristin will form a precipitate.

Filter off the precipitate using suction, and then wash the crystals with 25 mL of cold 95% ethanol.

The crystals are colorless.

Weigh the crystals

Record results _____ g.

Place some crystals in a melting point apparatus

Record results _____°C

## SELECTED REFERENCES—LIPIDS

Goodwin, T. W. *Biochemistry of Lipids.* Johns Hopkins, 1974.
Gunstone, F. *Lipid Chemistry.* Aspen, 2000.
Gunstone, F. *The Lipid Handbook.* Chapman Hall, 1986.
Gunstone, F. *Fatty Acid and Lipid Chemistry.* Aspen, 1999.
Gunstone, F. *Chemistry of Oils and Fats.* CRC, 2009.
Gurr, M. *Lipid Biochemistry.* Blackwell, 2002.
Hamilton, R. *The Chemistry and Technology of Oils and Fats.* Blackwell, 2007.
Harrow, B. *Textbook of Biochemistry.* Saunders, 1943.
Hitchcock, C. *Plant Lipid Biochemistry.* Academic, 1971.

Imai, Y. *Horizons in Lipid Biochemistry*. Hokkaido, 1984.
Koch, M. *Practical Methods of Biochemistry*. Wilkins, 1941.
Lawson, H. W. *Foods, Oils, and Fats*. Springer, 1995.
Mead, J. *Lipids: Chemistry, Biochemistry, and Nutrition*. Blackwell, 2007.
Paoletti, R. *Lipids* (3 volumes). Raven Press, 1975.
Smith, E. L. *Principles of Biochemistry*. McGraw-Hill, 1983.
Timberlake, K. *General, Organic, and Biological Chemistry*. Prentice-Hall, 2007.
Tyler, V. *Experimental Pharmacognosy*. Burgess, 1960.

# 3 Proteins

## PROTEINS

Proteins are essential constituents of all protoplasm and are essential food constituents. They are characterized by the fact that on hydrolysis they yield several dozen types of amino acids. Difference between proteins are largely a matter of number, and the kind of arrangement of such amino acids within the protein molecule. Proteins may be classified as

1. Simple, which includes albumins, globulins, glutelins, albuminoids, prolamines, histones, and protamines.
2. Conjugated, which includes nucleoproteins, glucoproteins, phosphoproteins, chromoproteins, and lipoproteins.
3. Derived, which includes proteans, metaproteins, coagulated proteins, proteoses, peptones, and peptides.

Medicinal products in this category include countless types of amino acid tablets, aminocaproic acid used as a blood-clotting agent, levodopa used as an antiparkinsonian agent, and the antiviral agents classified as interferon.

Experiments in this section include color reactions of various proteins, some precipitations, and the isolation of protein products from food and vegetable sources.

## EXPERIMENT 16: COLOR REACTIONS OF PROTEINS

Proteins belong to a group of one of the most complex of chemical substances. They are essential constituents of all protoplasm as well as an essential food constituent. Their complexity defies any one definitive classification system; however, the following one may prove useful to students beginning their study of this group:

1. Simple proteins
   Albumins are soluble in water and in dilute salt solutions, including egg white and serum albumin.
   Globulins are insoluble in water but soluble in neutral solutions of strong acids and bases and include egg yolk, fibrinogen of blood, and myosin of muscle.
   Glutelins are insoluble in all neutral solvents but soluble in dilute solutions of acids and bases and include glutenin in wheat and oryzenin in rice.
   Prolamins are insoluble in water but soluble in 70% alcohol and include gliadin of wheat and zein of maize.
   Albuminoids are insoluble in all neutral solvents and include keratin of horns and hair and the elastin of ligaments.

Histones are soluble in water and dilute acids but insoluble in dilute $NH_3$ and include globin from blood hemoglobin.

Protamines are soluble in $NH_3$ and water, and uncoagulable by heat; they are present in salmon and sturgeon sperm.

2. Conjugated proteins

Nucleoproteins, on hydrolysis, yield protein and nuclein and are present in glandular tissue and yeast.

Glycoproteins are proteins combined with a carbohydrate group and include mucin of saliva.

Phosphorproteins contain phosphoric acid combined with amino groups and include casein of milk and vitellin of egg yolk.

Chromoproteins are protein molecules combined with hematin or another colored substance. Examples are hematin of blood and phycocyan of seaweed.

Lecithoproteins are proteins combined with lecithin and are found in tissue fibrinogen.

3. Derived proteins

Proteans are products of enzymatic digestion of globulins, soluble in weak acids and bases, including edestan and myosin.

Metaproteins are soluble in weak acids and bases but insoluble in neutral water and include acid albuminate and alkali albuminate.

Coagulated proteins are water insoluble, resulting from action of heat, alcohol, heavy metal salts, and x-rays of proteins.

Proteoses are soluble in water and not coagulable by heat; they are precipitated by solutions of ammonium sulfate.

Peptones are soluble in water and are not precipitated by ammonium sulfate and are not coagulable by heat.

Peptides are combinations of two or more amino acids joined by the peptide linkage.

Proteins are complex substances of high molecular weight that, on hydrolysis, yield amino acids in addition to other products of decomposition, the character and quality of which vary with different kinds of proteins. The hydrolytic changes are produced by acids, bases, water, and specific enzymes. Proteins are polymers of amino acids just as polysaccharides are polymers of monosaccharides. The linking unit between terminals of amino acids is called a *peptide bond*:

$$^{+}H_3 N \underset{R}{C}HCO(NH \underset{R}{C}HCO)_n NH \underset{R}{C}HCOO^{-}$$

## Biuret Reaction

To 2 mL of a protein solution add 2 mL of a 10% NaOH solution and 1 drop of a 0.1% $CuSO_4$ solution. Mix well, and in case a pink to purple color has not yet developed, add another drop of 0.1% $CuSO_4$ solution and continue until you get some color.

Be careful not to put in excess copper, since an excess combined with the protein substances obscures the true colors.

Repeat the same test on 2% solutions of album in and peptone and rank the intensity of the color reactions.

Results:

1. _____
2. _____
3. _____

## NINHYDRIN REACTION

Ninhydrin, which chemically is triketohydrindene hydrate, reacts very delicately in detecting proteins and amino acids but is also very sensitive to pH. The colors developed differ with the character of the protein, the amino acids, or a mixture thereof. The structure of ninhydrin is

Make up 4 mL of a protein solution and titrate accordingly in order to achieve neutrality. It is crucial for obtaining accurate results that the solution be at pH 7. Add 1 mL of a 0.1% ninhydrin solution. Mix and boil for 1 minute, then set aside to cool. Repeat the same procedure with a 0.2 M casein solution and a 0.1 M leucine solution, and use water in the fourth test tube as a standard. Record the results.

Casein _____
Leucine _____
Water _____

## XANTHOPROTEIC REACTION

This reaction involves the nitration of the benzene nucleus and resulting colors when the solution is rendered alkaline. Remember, the protein must have an amino acid group with an aromatic nucleus and not all amino acids have this nucleus.

To a few milligrams of tyrosine, tryptophan, glycine, and phenylalanine in four small test tubes, add 1 mL of concentrated $HNO_3$ to each tube and boil until the amino acids are dissolved. Now cool and add a few drops of 40% NaOH solution until the solution is slightly alkaline as determined by litmus paper. Record the results.

Tyrosine _____
Tryptophan _____

Glycine      _____

Phenylalanine _____

## MILLON'S REACTION

This reaction is given by most of the monohydroxy benzene derivatives and, to some extent, by certain dihydroxy benzene derivatives.

To 2 mL of solutions of tyrosine, resorcinol, and tryptophan, add several drops of Millon's Reagent and bring to boiling in a water bath. If no color appears at first, continue to add drops of the Millon's Reagent. Record the results.

Tyrosine      _____

Resorcinol   _____

Tryptophan _____

## EHRLICH DIAZO REACTION

This reaction involves the interactions of diazo-benzene-sulfanilic acid with phenol imidazole groups. The reaction is not only useful for qualitative studies, but it has been developed into quantitative methods for related substances.

To 3 mL of a 0.5% solution of sulfanilic acid in 2% HCl, add an equal volume of a 0.5% solution of $NaNO_2$. After shaking and letting stand for 1 minute, add 1 mL of this solution to 10 test tubes containing tyrosine, histidine, epinephrine, and bile pigments. Mix each tube well and then add 10% $Na_2CO_3$ solution until each solution is now alkaline. Record the results.

Tyrosine      _____

Histidine     _____

Epinephrine   _____

Bile pigments _____

## EXPERIMENT 17: PROTEIN PRECIPITATIONS

Proteins act as weak acids and weak bases in equilibrium with each other. If positively charged, ions are introduced, proteinates are formed, and since the metal proteinates are only sparingly soluble, the protein is precipitated as a salt.

### PRECIPITATION BY POSITIVE IONS

Prepare a fresh solution of 10% fresh beef blood serum and add 5 mL of this solution to each of four test tubes. Add 5 mL of a 1% solution of $ZnSO_4$ to the first, 1% $CdCl_2$ to the second, 1% $CuSO_4$ to the third, and 1% $HgCl_2$ to the fourth. In tubes 3 and 4, also add 0.5 mL of a. 02 N NaOH solution and record the results.

$ZnSO_4$ _____

$CdCl_2$   _____

$CuSO_4$ _____
$HgCl_2$ _____

Explain the differences in reactions. (Hint: consider the effect of hydrogen ion concentration on protein precipitation.)

Determine the pH of the filtrates of each tube using bromphenol blue and bromocresol purple. Record the results.

$ZnSO_4$ filtrate pH _____
$CdCl_2$ filtrate pH _____
$CuSO_4$ filtrate pH _____
$HgCl_2$ filtrate pH _____

## PRECIPITATION BY ANIONS

Prepare four tubes of 5 mL fresh beef blood serum (10%) as in the previous experiment. To the first tube, add 1 mL of a 0.7 N $H_2SO_4$ and 1 mL of a 10% sodium tungstate solution. In tube 2, substitute 0.5 mL of a 10% sodium ferrocyanide solution; in tube 3, change to 1 mL of a 1% sodium phosphomolybdate solution, and in tube 4 use 2 mL of a 1.2% sodium picrate solution, but do not use any $H_2SO_4$ in this tube. Record the results.

Sodium tungstate _____
Sodium ferrocyanide _____
Sodium phosphomolybdate _____
Sodium picrate _____
Explain the differences in results obtained:
Determine the pH of the filtrates from each tube using bromphenol blue and
    bromcresol purple. Record the results:
Sodium tungstate filtrate pH _____
Sodium ferrocyanide filtrate pH _____
Sodium phosphomolybdate filtrate pH _____
Sodium picrate filtrate pH _____

Explain the differences in pH results in the precipitation of proteins using cations and anions:

## EXPERIMENT 18: PREPARATION OF CASEIN FROM SKIMMED MILK

The proteins of milk include casein, lactalbumin, and lactoglobulin, with casein comprising about 80% of the total protein. Casein is a phosphoprotein yielding phosphoric acid and amino acids on hydrolysis. The isoelectric point of casein is a pH of about 4.6, but the pH of milk is about 7.0, thus indicating that the casein in milk is in alkaline combination, probably in the form of calcium casein.

Casein is insoluble in water. The addition of acid to milk serves to precipitate it. The casein precipitated can be redissolved in an alkaline solution and then reprecipitated by acids, thus serving as a method of purifying the protein.

$$Na\ caseinate + HCl \rightarrow NaCl + H\ caseinate\ (or\ casein)$$

$$Casein\ hydrate + HCl \rightarrow casein\ HCl + H_2O$$

## THE PREPARATION PROCESS

To 100 mL of skimmed milk in a 1000 mL beaker, add 300 mL of $H_2O$.[*] Mix well and remove 40 mL of the mixture to a small beaker and add, carefully, drop by drop a 10% $H_2SO_4$ solution of a coarsely flocculent precipitate of casein. Now return the 40 mL to the larger sample of diluted milk and, in the same way, drop by drop add 10% $H_2SO_4$ in the same ratio while constantly stirring.

Be very careful not to add the acid too fast, as you miss the critical point and the casein will redissolve to form casein sulfate. If this should happen, you may add some 10% NaOH solution until the casein reprecipitates. To the crude casein add, 30 mL of water and stir until the precipitate is finely divided. To this suspension, now add gradually a 10% solution of NaOH and stir until all the precipitate has now dissolved.

The next step is to reprecipitate the casein by slowly adding a 10% acetic acid solution. After achieving a good, coarse flocculent precipitate, decant the supernatant and wash the precipitate with 300 mL of distilled water several times.

Finally, transfer the precipitate to a Buchner funnel, filter, and suction dry by vacuum. Remove the casein from the funnel, place in a mortar, and gradually add 50 mL of ethanol while kneading the powder. Again filter by suction. Repeat this process two more times and finally dry.

After the casein is thoroughly dry, enter your results below:

Weight of 1000 mL beaker and 100 mL milk _____
Weight of 1000 mL beaker _____ (minus) _____
Weight of 100 mL milk _____
Weight of casein precipitate _____

Percentage composition of casein

$$\frac{Weight\ of\ casein}{Weight\ of\ milk} \times 100 = \underline{\hspace{2cm}}\%$$

## EXPERIMENT 19: PREPARATION OF EDESTIN FROM HEMP SEED

Edestin is a protein globulin present in hemp, commonly called marijuana, or botanically known as *Cannabis sativa*. The seeds contain some fats that must be extracted by the use of a suitable fat solvent.

To do this, begin by thoroughly grinding the seed in a mortar[†] then transfer to a small flask and cover with a layer of benzene or naphtha. Stopper the flask and place on a magnetic stirrer to agitate overnight. Filter this by gravity, dispose of the filtrate,

---

[*] Be sure to weigh beaker empty and with 100 mL milk and record results.
[†] Be sure to weigh and record weight or hemp seeds prior to adding the solvent.

and repeat the procedure with the seeds. Then expose the seeds and solvent on a watch glass and allow all solvent to evaporate. Caution: do not use any flames.

To 10 g of the finely ground fat-free hemp, add 100 mL of a 1% NaCl solution in a 250 mL flask. Place in a water bath maintained at 60°C for 1 hour. Do not allow the temperature to go above this number as it will cause the edestin to coagulate. Now filter the warm solution until you have collected about 75 mL of filtrate. This should be slightly turbid due to the separation of the edestin. Warm this filtrate in a water bath again at 60°C until clear.

Once the solution is clear, discontinue heating and allow the flask to remain in the water overnight. This effects a slow cooling process, which results in the formation of larger crystals of edestin being formed than if the cooling were done more rapidly.

Gravity filler the edestin crystals and perform percentage calculations as follows:

$$\frac{\text{Weight of edestin obtained}}{\text{Weight of hemp seed}} \times 100 = \underline{\qquad} \%$$

## EXPERIMENT 20: PAPER CHROMATOGRAPHY OF AMINO ACIDS

Proteins can be hydrolyzed in acidic or basic solution, or with enzymes causing the peptide bonds to give shorter polypeptides, which in turn are further degraded to amino acids. Total hydrolysis of proteins can be achieved in 20% HCl at 100°C for 12 to 48 hours. However, adequate hydrolysis for purposes of this experiment can be achieved in less than 1 hour.

$$-NH-\overset{\overset{R}{|}}{C}H-\overset{\overset{O}{\|}}{C}-NH-\overset{\overset{R}{|}}{C}H-\overset{\overset{O}{\|}}{C}-NH-\overset{\overset{R}{|}}{C}H-\overset{\overset{O}{\|}}{C}$$
(Protein)

$$\rightarrow n(NH_2-\overset{\overset{R}{|}}{C}H\ COOH)$$
(Amino acids)

These hydrolysates can be analyzed for their amino acid content by methods such as paper chromatography. Some of the more common amino acids are listed with their approximate $R_f$ values and their percentage composition in casein, gelatin, and hair.

|  |  | % | | |
| --- | --- | --- | --- | --- |
| **Amino Acid** | **$R_f$ Value** | **Casein** | **Gelatin** | **Hair** |
| Aspartic acid | .32 | 6.8 | 6.7 | 3.9 |
| Glutamic acid | .40 | 2.6 | 25.5 | 4.1 |
| Threonine | .51 | 4.7 | 1.9 | 8.5 |
| Alanine | .59 | 2.9 | 8.7 | 2.8 |
| Tyrosine | .62 | 6.1 | 0.4 | 2.2 |
| Valine | .75 | 6.9 | 2.5 | 5.5 |
| Leucine | .79 | 8.8 | 4.6 | 11.2 |
| Phenylalanine | .82 | 4.8 | 2.2 | 2.4 |
| Proline | .85 | 10.9 | 18.0 | 4.3 |

## PAPER CHROMATOGRAPHY PROCEDURE

Using a 24 × 15.5 cm rectangle of filter paper (be careful not to touch anything but the edges of the paper or a false chromatogram may be given), pencil a line across the bottom about 2 cm from the edge. Then about 2 cm apart, make pencil dots for 10 equally spaced places where you will apply the various solutions. In the first 7 spots, you will pipet a dot 1–2 mm in diameter of 0.1 M solutions of aspartic acid, glutamic acid, threonine, alanine, valine, leucine, and proline, and in the last three spots you will pipet 1–2 mm diameter dots of the hydrolysates of casein, gelatin, and hair. Each of these solutions should have been acidified with 10 drops of 19% HCl for each 10 mL of solution.

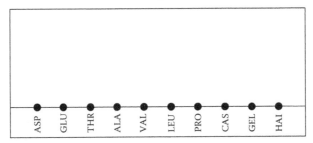

Once you have made the 10 spots and they have dried, repeat the pipet procedure to make sure you have sufficient concentration of each substance.

Once the spots have dried lay the chromatogram on a clean sheet of paper and prepare a developing chamber by placing an 80% solution of phenol ($C_6H_5OH$) on the bottom of a 2000 mL beaker, making sure that it does not become more than 1.5 cm in depth, thereby not touching your amino acid spots on the chromatogram. Now place the chromatogram down in the phenol and cover the beaker with a lid. Allow the chromatogram to develop, which should take a couple of hours.

When the phenol front has reached close to the top of the filter paper, you should remove it with tweezers and dry thoroughly. It may take a couple of hours for the solvent to evaporate, so the spraying session may be reserved for the next lab period. Spray the paper uniformly with a ninhydrin spray and place the paper in a 110°C oven. After 5 minutes, remove the paper and observe the colored spots. Calculate $R_f$ values for all 10 substances and record:

|  | $R_f$ Value |
|---|---|
| Aspartic acid | _____ |
| Glutamic acid | _____ |
| Threonine | _____ |
| Alanine | _____ |
| Valine | _____ |
| Leucine | _____ |
| Proline | _____ |
| Casein | _____ |
| Gelatin | _____ |
| Hair | _____ |

On the last three, you may want to compare results with the table at the beginning of the experiment.

## SELECTED REFERENCES—PROTEINS

Anffinsen, C. *Advances in Protein Chemistry*. Academic, 1988.
Bohak, Z. *Biochemical Applications of Proteins*. Academic, 1977.
Branden, C. *Introduction to Protein Synthesis*. Garland, 2000.
Crabb, J. *Techniques in Protein Chemistry*. Academic, 1995.
Creighton, T. *Proteins*. Freeman, 1984.
Doonan, S. Peptides and Proteins. Cambridge, 2002.
Howard, G. *Modern Protein Chemistry*. CRC Press, 2001.
Kyte, J. *Structure in Protein Chemistry*. Garland, 2006.
Magnusson, S. *Proreolytic Enzymes*. Pergamon Press, 1978.
Makowski, G. *Clinical Chemistry*. Elsevier, 2012.
Neurath, H. *Proteins* (4 volumes). Academic, 1979.
Schulze, G. E. *Principles of Protein Structure*. Springer, 1979.
Scopes, R. K. *Protein Purification*. Springer, 1989.
Vullafranca, J. *Techniques in Protein Chemistry*. Academic, 1991.

# 4 Enzymes

## ENZYMES

Enzymes are chemically proteins but have special catalytic properties to propel reactions in biological systems. By general agreement, their names end in "ase." One of several classifications of enzymes follows:

1. Esterases
2. Proteinases
3. Amidases
4. Phosphorylases
5. Carbohydrases
6. Oxidases

Enzymes are, for the most part, soluble in water and in dilute alcohol. They are temperature sensitive and are most active in the body (37°C–38°C temperatures). Their optimum pH activity is near neutrality (7.0).

Some medicinal products using enzymes include chymotrypsin in cataract surgery, and plant protease enzymes used in the treatment of soft tissue inflammation. Also sutilain is a proteolytic enzyme used to dissolve necrotic tissue occurring in second- and third-degree burns, and hyaluronidase used as a local anesthetic in surgery of the eye.

Experiments in this section illustrate enzyme activity in the digestive processes, including the salivary, gastric, and intestinal areas.

## EXPERIMENT 21: ENZYME ACTIVITY IN SALIVARY DIGESTION

Enzymes are catalysts produced as a result of cellular activity that occurs in abundance in plant and animal tissue. Although a number of enzymes have been isolated, many have been identified by their behavior toward substrates. By general agreement, most enzymes end in "ase," with the exception of a few that were well known prior to the establishment of the nomenclature by the International Union of Biochemistry (IUB). Chemically, they are proteins and are classified in six major categories. A brief outline is

1. Oxidoreductases, also known as dehydrogenases or oxidases, include alcohol dehydrogenases and glutamic dehydrogenase in the liver.
2. Transferases, which catalyze the transfer of one-carbon groups, include aldehydes, ketones, phosphorus, and sulfur groups.
3. Hydrolases catalyze hydrolysis of ester, ether, peptide, and other bonds, including such examples as pepsin and rennin.

4. Lyases catalyze removal of groups from substrates by means other than hydrolysis, leaving double bonds. They include ketone, aldehyde, and carbon–oxygen Iyases.

5. Isomerases, including enzymes, catalyze interconversion of optical and geometric isomers.

6. Ligases catalyze the linking together of two compounds coupled to the breaking of a pyrophosphate bond. They include Succinate Co-A and Acetyl Co-A.

The series of exercises in this experiment will serve to identify enzyme activity of body fluids. One of these is the classic Wolgemuth method used to determine the erythrodextrin reaction with iodine. Suspend 2 g of soluble starch in 25 mL of water. Gradually heat to boiling while stirring and continue until solution is nearly clear. Cool to room temperature and dilute to 100 mL. Preserve with a few mL of toluene. Next dilute 1 mL of 0.1 N $I_2$ solution to 120 mL, which gives you a N/1200 $I_2$ solution.

Label seven test tubes 1 through 7 (16 × 150 mm) and measure off 2.5, 2.0, 1.6, 1.2, 0.8, 0.4, and 0.2 mL of fresh saliva. To these now add, in the same order, 0.0, 0.5, 0.9, 1.3, 1.7, 2.1, and 2.3 mL of distilled water to make equal volumes of 2.5 mL. Shake well to mix and then rapidly add 2.5 mL of a 2% soluble starch solution. Shake each tube well and place in a water bath at 40°C. Keep tubes in the bath for 30 minutes, then at once immerse in ice water and add 2 mL of N/1200 $I_2$ solution to each. Shake each tube well, and note the minimum concentration of saliva at which the erythrodextrin reaction persists.

From your data calculate the number of millilitre of 2% starch that 1 mL of your saliva will hydrolyze to the achromic point in 30 minutes at 40°C.

Calculations:

The second activity of this experiment will show the presence of nitrites ($NO_2^-$) in saliva. To 1 mL of saliva, add 2 drops of 10% $H_2SO_4$. Mix well and add 2 drops of a 10% KI solution and a drop of starch solution. The nitrous acid ($HNO_2$) formed liberates the iodine that is detected by the starch.

Record the color reaction: _____.

The third activity shows the presence of sulfocyanates in saliva. To 1 mL of saliva, add 1 drop of 2% $FeCl_3$ solution and acidify slightly with HCl. If a pink to red color develops, it is due to the sulfocyanates forming a complex ion with the ferric ions or to a precipitate of ferric phosphate. If coloration is due to sulfocyanates, 2 to 5 drops of 1% $HgCl_2$ solution will convert the colored compound to colorless mercuric sulfocyanate.

Record results:

## EXPERIMENT 22: ENZYME ACTIVITY IN GASTRIC DIGESTION

Pepsin is one of the major enzymes involved in gastric digestion and has a precursor called pepsinogen. Evidence for the existence of the two forms is based on the difference in behavior of the two toward acids. Alkalies destroy pepsin, whereas acids protect and aid it. Alkalies protect pepsinogen, whereas acids destroy it by converting it into pepsin.

In this experiment, take 10 g of well-washed gastric mucous membrane, add about 15 g of sea sand, and grind in a mortar. Gradually add to this 100 mL of water containing two drops of a 10% $Na_2CO_3$ solution. Mix well and strain through cheesecloth. Next, measure off four 5 mL portions of this extract into 25 × 200 mm test tubes, and label them 1 through 4.

To tube 1 add 13.4 mL of water and 1.6 mL of a 1 N NaCl solution and mix well. You now have 20 mL of a solution containing 1.6 mL of NaCl but which is neutral and has not been acted on by acids or bases. Since there is no acid medium, no pepsin activity is evident. To tube 2, add 11.4 mL of water, 2 mL of 1 N HCl, and 1.6 mL of 1 N NaCl and mix well. Here, we have 20 mL of solution that is 0.1 N HCl and contains 1.6 mL of 1 N NaCl. This solution should digest protein readily, but it does not tell us whether the original extract contains pepsin, pepsinogen, or both.

To tube 3, add 9.4 mL of water and 1.6 mL of a 1 N $Na_2CO_3$ solution and mix well. This gives you 16 mL of a 0.1 N $Na_2CO_3$ solution. If pepsin is present in the original extract, it should be destroyed, but the pepsinogen is not destroyed. Next, add 3.6 mL of 1 N HCl and 0.4 mL of water. You now have 20 mL of a 0.1 N HCl solution again containing 1.6 mL of 1 N NaCl. In case pepsin is present, you have favorable conditions for its action on proteins.

To tube 4, add 0.6 mL of 1 N HCl and 0.4 mL of water and mix well. If pepsinogen is present, it should be converted to pepsin. Next, add 1.6 mL of a 1 N $Na_2CO_3$ solution and 2.4 mL of water and mix well. You now have 10 mL of a 0.1 N $Na_2CO_3$ solution. If pepsinogen is still present, it will not be destroyed by the $Na_2CO_3$, but if the pepsinogen has previously been converted to pepsin, it will now be destroyed by the base. Next, add 3 mL of 1 N HCl and 7 mL of water and shake. The final solution is of a 0.1 N HCl solution containing 1.6 mL of 1 N NaCl. This last addition of acid is made in order to render the solution acid so that pepsin can act if present.

Once having done all this you are now ready to proceed to the final step and begin recording observations. Introduce 0.5 g of dried egg white into each of the four tubes. Mix well and incubate in a water bath at 45°C for 60 minutes. Agitate the tubes frequently and observe the relative rates of solution in the four tubes. The differences in rates of digestion should be quite distinct. (Hint: two of the four tubes should show some real enzyme activity.)

Record all observations:

The second portion of the gastric enzymes experiment involves rennin, another enzyme of great importance in gastric digestion. Its efficacy depends on the presence of calcium salts for precipitation of end products. To illustrate this action, prepare four 25 × 200 mm test tubes as shown by the following table:

| Tube | Milk | Additions |
|------|------|-----------|
| 1. | 10 mL | 2 mL of water |
| 2. | 10 mL | 1 mL 1 N $K_2C_2O_4$ and 1 mL water |
| 3. | 10 mL | 1.25 mL 1 N $CaCl_2$ and 1 mL $K_2C_2O_4$ |
| 4. | 0 | 1 mL 1 N $K_2C_2O_4$ and 1 mL water |

Set the four tubes in a water bath at 40°C and add 1 mL of a rennin solution prepared by grinding 5 mL of rennin with sand and adding 25 mL of water. Record the results of curdling action after 15 minutes have elapsed.

Results:

## EXPERIMENT 23: ENZYME ACTIVITY AND INTESTINAL DIGESTION

Pancreatin or pancreatic lipase acts in the body to metabolize fatty substances. To illustrate this action, prepare 100 mL of a 10% solution of pancreatin. Allow it to stand for 24 hours and strain any residue through cheesecloth. Then add 1 drop of toluene and allow it to digest overnight at 40°C.

Now prepare nine test tubes according to the table below, using olive oil or another suitable fatty acid substance:

| Tube | Olive Oil (mL) | Water (mL) | Others | Enzyme Solution (mL) |
|------|----------------|------------|--------|----------------------|
| 1 | 0.5 | 4.5 | 0 | 5.0 |
| 2 | 0.5 | 4.5 | 0 | 5.0 (boil 5 min) |
| 3 | 0 | 4.5 | 0 | 5.0 |
| 4 | 0 | 9.5 | 0 | 0 |
| 5 | 0.5 | 9.5 | 0 | 0 |
| 6 | 0.5 | 8.5 | 1 mL bile | 0 |
| 7 | 0.5 | 3.5 | 1 mL bile | 5.0 |
| 8 | 0 | 8.5 | 1 mL bile | 0 |
| 9 | 0 | 3.5 | 1 mL bile | 5.0 |

Record the results:
Tube
1.
2.
3.
4.
5.
6.
7.
8.
9.

## EXPERIMENT 24: BILE PIGMENTS

The two most common bile pigments are bilirubin and biliverdin. Bilirubin is an isomer of hematoporphyrin, the iron-free derivative obtained from hematin by the action of strong acids. When oxidized, bilirubin yields various stages of oxidation from the blue bilicyanin to the green biliverdin to the yellow choletelin. The following tests are based on the conversion of bilirubin into bilicyanin and biliverdin.

Structure of Bilirubin

Begin the lab session by performing Gmelin's test, which uses 5 mL of concentrated $HNO_3$ to which 5 mL of a 10% bile solution is added. Set aside for 30 minutes and record color reactions.

Observations:

Another bile pigment test is the Huppert–Steensma. Mix 10 mL of a 10% bile solution with an equal volume of lime water. Shake well. The bile pigments are precipitated as calcium salts. Remove the precipitate and dissolve it in 5 mL of 95% $C_2H_5OH$ containing 0.5 mL of concentrated HCl. Add a few drops of a 0.5% $NaNO_2$ solution and warm the mixture in a water bath. Add a few drops of concentrated HCl. The presence of biliverdin is indicated by a color reaction. Record your observations.

Results:

Still another test for bile pigments is the Hammarsten test. This reagent is made by the addition of 5 mL of 0.25 M $HNO_3$ to 95 mL of 0.25 M HCl. Mix well and allow to cool. Now to 1 mL of this reagent add 4 mL of 95% $C_2H_5OH$ and in a porcelain dish add a few drops of the 10% bile solution. If sufficiency bilirubin or biliverdin are present, a more or less permanent green color develops at once. A yellow is color indicates the development of choletelin.

Results:

## SELECTED REFERENCES—ENZYMES

Abelson, J. N. *Methods in Enzymology*. Academic Press, 1988.
Bergmeier, H. *Methods of Enzymatic Analysis*. Academic, 1983.
Bugg, T. *Introduction to Enzyme and CoEnzyme Chemistry*. Wiley, 1997.
Chaplin, M. *Enzyme Technology*. Cambridge University Press, 1990.
Henry, R. J. *Clinical Chemistry*. Hoeber, 1964.
Martin, G. J. *Clinical Enzymology*. Little, Brown, 1958.
Palmer, T. *Enzymes: Biochemistry, Biotechnology, and Clinical Chemistry*. Hopwood, 2001.
Suckling, C. J. *Enzyme Chemistry*. Springer, 1998.
Von Eiler, H. *General Chemistry of the Enzymes*. Forgotten, 2012.
Wilson, K. *Principles and Technology of Protein Biochemistry*. Cambridge University Press, 2000.
Wong, C. *Enzymes in Synthetic Organic Chemistry*. Academic Press, 1994.

# 5 Inorganics

## INORGANICS

Inorganics refers to chemical and pharmaceutical products in which the carbon atom is not present, or if present, is not the central unit of the structure. Essentially, this could mean any other element in the periodic chart. Pharmaceutically, this includes two categories:

1. Binary compounds
2. Acids, bases, and salts.

The most common inorganics are hydrogen and oxygen, which comprise the water covering about three quarters of the world's surface.

Medicinal products with an inorganic basis include such common products as tincture of iodine, used as an antibacterial; Epsom salts (magnesium sulfate), sometimes employed as a laxative; hydrogen peroxide, used as an antiseptic; and magnesium and aluminum hydroxides used an antacids.

Experiments in this section include acid-base chemistry, assays of silver compounds, and the isolation of the element bismuth from a popular commercial antacid preparation.

## EXPERIMENT 25: ACIDIMETRY AND ALKALIMETRY IN INORGANIC ANALYSIS

Acidimetry is the measurement of the quantity of acid in a given sample by titration with a suitable alkali. Alkalimetry is the measurement of the quantity of alkali in a given sample by titration with a suitable acid. In both cases indicators are used to identify end points in titration. In each case, the principle involved is the same; namely, the addition of a chemically equivalent amount of standard solution of an acid or a base to the sample of base or acid, respectively, being assayed.

*Procedure for the assay of boric acid ($H_3BO_3$).* Dry 2 g of boric acid powder over sulfuric acid in a dessicator for 5 hours prior to performing this experiment. Then place the powder in 100 mL of a mixture of glycerine and water that has been previously neutralized using phenolphthalein T.S. Add phenolphthalein T.S, and titrate with 1.0 M NaOH. Now discharge the pink color by the addition of 50 mL of glycerine, then continue the titration with NaOH until the pink color reappears. Each milliliter of the 1.0 M NaOH is equivalent to 61.84 mg of $H_3BO_3$.

Record results:

_____ mL of NaOH × 61.84 = _____ mg $H_3BO_3$

Percentage Assay $\dfrac{x}{2\ g} \times 100 =$ _____ %

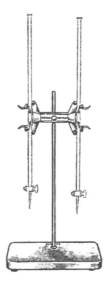

*Procedure for the assay of methenamine $C_6H_{12}N_4$ (hexamethylenetetramine).* Place i g of methenamine (previously dried over $H_2SO_4$ for 4 hours) in a 100 mL beaker and add 40 mL of 1.0 N $H_2SO_4$; boil gently. Add water from time to time, if necessary, until the odor of formaldehyde is gone. Cool and add 20 mL of water plus methyl red T.S. and titrate the excess with 1.0 N NaOH. Each milliliter of the 1.0 N $H_2SO_4$ is equivalent to 35.05 mg of $C_6H_{12}N_4$. Record results:

_____ mL $H_2SO_4$ × 35.05 = _____ mg $C_6H_{12}N_4$.

## EXPERIMENT 26: THE ASSAY OF SILVER IN STRONG SILVER PROTEIN

Compounds containing silver and mercury ions can be readily converted into soluble Ag or $Hg^2$ salts, respectively, and may be estimated by direct titration with standard ammonium thiocyanate using ferric ammonium sulfate as the indicator. The method is based on the quantitative precipitation of the corresponding thiocyanate. Thus:

$$AgNO_3 + Nh_4SCN \rightarrow AgSCN + NH_4NO_3 \text{ (for silver)}$$

$$Hg(NO_3)_2 + 2NH_4SCN \rightarrow Hg(SCN)_2 + 2NH_4NO_3 \text{ (for mercury)}$$

In each case, when all the metallic ion has been precipitated as thiocyanate, the $NH_4SCN$ reacts with the ferric ion indicator to form red-colored ferric thiocyanate marking the end point of titration.

## TABLE 5.1
## A Comparison of [H₃O⁺], [OH⁻], and Corresponding pH Values at 25°C

| pH | $[H_3O^+]$ | $[OH^-]$ | |
|----|------------|----------|---|
| 0 | $10^0$ | $10^{-14}$ | |
| 1 | $10^{-1}$ | $10^{-13}$ | |
| 2 | $10^{-2}$ | $10^{-12}$ | |
| 3 | $10^{-3}$ | $10^{-11}$ | Acidic |
| 4 | $10^{-4}$ | $10^{-10}$ | |
| 5 | $10^{-5}$ | $10^{-9}$ | |
| 6 | $10^{-6}$ | $10^{-8}$ | |
| 7 | $10^{-7}$ | $10^{-7}$ | Neutral |
| 8 | $10^{-8}$ | $10^{-6}$ | |
| 9 | $10^{-9}$ | $10^{-5}$ | |
| 10 | $10^{-10}$ | $10^{-4}$ | |
| 11 | $10^{-11}$ | $10^{-3}$ | Basic |
| 12 | $10^{-12}$ | $10^{-2}$ | |
| 13 | $10^{-13}$ | $10^{-1}$ | |
| 14 | $10^{-14}$ | $10^0$ | |

## Acidimetric Titrations Table

| Substance | Indicator | Equiv mL/mg | Req. % |
|-----------|-----------|-------------|--------|
| Benzoic acid | Phenolphthalein | 12.21 | 99.3 |
| Citric acid | Phenolphthalein | 64.04 | 99.5 |
| Saccharin Na | Phenolphthalein | 13.81 | 99.5 |
| Citrated caffeine | Phenolphthalein | 6.404 | 48–52 |
| Tartaric acid | Phenolphthalein | 75.04 | 99.7 |
| Vinobarbital | Thymol blue | 22.43 | 98.5 |

## Alkalimetric Titrations Table

| Substance | Indicator | Equiv. mL/mg | Req. % |
|-----------|-----------|--------------|--------|
| Calamine | Methyl red | 40.69 | 98 |
| Dimenhydrinate | Methyl red | 25.54 | 53–55 |
| Meperidine HCl | Methyl red | 5.676 | 95 |
| Methadone HCl | Methyl red | 6.918 | 93 |
| Methamphetamine | Methyl red | 9.285 | 90 |
| Nicotinamide | Methyl red | 12.21 | 98 |
| Lidocaine | Bromcresol blue and Methyl red | 23.43 | 99 |
| Saccharin Na | Methyl red | 12.06 | 95 |

*Procedure for the assay of strong silver protein.* Ignite 2 g of strong silver protein in a porcelain crucible until all the carbon is burned off. Cool the crucible and place it in a 400 mL beaker and fill it with dilute $HNO_3$ (0.5 M). When the reaction has subsided, add water to immerse the crucible, cover the crucible, and heat on a water bath until all the silver has dissolved. Next, remove and rinse the crucible and filter the solution into an Erlenmeyer flask. Wash the filter and residue thoroughly with water and cool the filtrate. Add 2 mL of ferric ammonium sulfate T.S. and titrate with 0.1 M $NH_4SCN$. Each milliliter of the $NH_4SCN$ used is equivalent to 10.79 mg of silver.

$$Ag + 2HNO_3 \rightarrow AgNO_3 + H_2O + NO_2$$

$$Ag_2O + 2HNO_3 \rightarrow 2AgNO_3 + H_2O$$

Record results:

_____ mL $NH_4SCN \times 10.79$ _____ mg silver

## EXPERIMENT 27: ASSAY OF HYDROGEN PEROXIDE SOLUTION

Hydrogen peroxide is generally sold in drug stores as a 3% solution used as a topical antiseptic. Dental offices use a 30% solution for teeth-whitening procedures, and this strength also finds use in industry as a strong oxidizing agent. In this exercise, we will perform an assay for $H_2O_2$ and compare with the 3% label.

*Procedure for assay of hydrogen peroxide.* Measure 2 mL of $H_2O_2$ solution and transfer it to a 125 mL flask containing 20 mL water. Add 20 mL of dilute $H_2SO_4$ and titrate with 0.1 M $KMNO_4$. Each milliliter of 0.1 M $KMNO_4$ used is equivalent to 1.701 mg of $H_2O_2$. Record results:

_____ mL 0.1 M $KMNO_4 \times 1.701$ _____ mg $H_2O_2$

Although hydrogen peroxide generally acts as an oxidizing agent and is reduced to water, in this assay it reduces the potassium permanganate and is oxidized, yielding oxygen gas ($O_2$).

The equation for this reaction is

$$5H_2O_2 + 2KMNO_4 + 3H_2SO_4 \rightarrow 5O_2 + 2MNSO_4 + K_2SO_4 + 8H_2O$$

In order to comply with United States Pharmacopeia/National Formulary definition, a solution of hydrogen peroxide should contain not less than 2.5% and not more than 3.5% W/V of $H_2O_2$. Calculate the percentage of $H_2O_2$ in the sample assayed and compare the strength of the solution with the USP/NF requirement.

2.5% of 2 mL _____ mL $H_2O_2$
3.5% of 2 mL _____ mL $H_2O_2$

Compare your previous answer from the assay to these parameters and record results:

## EXPERIMENT 28: ISOLATING BISMUTH FROM PEPTO-BISMOL®

The popular medicine Pepto-Bismol has a long history of use in medicine as an OTC remedy for both upper and lower GI symptoms, including upset stomach and diarrhea. Each tablet contains 262 mg of bismuth subsalicylate plus inert ingredients, including benzoic acid, magnesium aluminum silicate, saccharin sodium, sorbic acid, and water.

The general formula for bismuth subsalicylate is $C_7H_5BiO_4$ giving it a molecular weight of 362 of which 209 is the element bismuth or about 57%.

In order to isolate the bismuth from the drug you need to reduce it chemically in somewhat the same way iron ore is reduced to make iron.

*Procedure*: Begin by weighing out 20 Pepto-Bismol tablets and record. Now calculate the weight of the drug bismuth salicylate by multiplying the 20 pills times 262 mg, which is the weight of the bismuth subsalicylate in each tablet. This information is available on the label of the box.

Begin by grinding the tablets into a powder using a mortar and pestle. Dissolve this powder in 200 mL of 2.0 M HCl. Filter the sludge solution using either a mechanical or electric vacuum pump. This should result in a clear pink solution containing mainly bismuth subsalicylate. The next step is to dip aluminum foil into the solution turning it black. The acid dissolves the aluminum, which reacts with the bismuth ions, which then precipitate out as metal particles.

Filter this mixture through cloth. The remaining filtrate is the bismuth metal. Heat this metal using a Bunsen burner, propane torch, or electric heater set on high, which should yield actual solid metal particles with the characteristic iridescence of bismuth.

Calculations:

a. Weight of the 20 tablets _____ mg
b. Weight of $C_7H_5BiO_4$ (20 × 262 mg) _____ mg
c. Theoretical weight of Bi (.57 × (b) _____ mg
d. Actual weight of bismuth obtained _____ mg
e. Percentage yield of the experiment _____ %

## SELECTED REFERENCES—INORGANICS

Block, J. *Inorganic, Medicinal, and Pharmaceutical Chemistry*. Lea & Febiger, 1984.
Bothara, K. J. *Inorganic Pharmaceutical Chemistry*. Pragati, 2008.
Collins, F. A. *Advanced Inorganic Chemistry*. Wiley, 1962.
Jenkins, G. L. *Quantitative Pharmaceutical Chemistry*. McGraw-Hill, 1957.
Kastura, A. V. *Pharmaceutical Inorganic Chemistry*. Pragmatic, 2008.
Moeller, T. *Inorganic Chemistry*. Wiley, 1952.
Parrott, E. *Pharmaceutical Chemistry*. Burgess, 1970.
Rogers, C. *Inorganic Pharmaceutical Chemistry*. Lea & Febiger, 2004.
Singh, A. *Pharmaceutical Inroganic Chemistry*. Delhi, 2008.

# 6 Vitamins

## VITAMINS

The word "vitamin" is a contraction of the term "vital amine." Vitamins constitute a total of 24 organic compounds that have been characterized as dietary essential. Many vitamins function biochemically as precursors in the synthesis of coenzymes necessary in human metabolism. There are two major categories of vitamins:

1. The lipid-soluble vitamins include A, D, E, and K. In addition to common solubility, they are usually associated with the lipids of foods and are absorbed in the intestine with these dietary lipids.
2. The water-soluble vitamins are B and C. The B complex refers to thiamine, riboflavin, pyridoxine, nicotinic acid, pantothenic acid biotin cyanocobalamin, and folic acid.

The medicinal products in this category include many vitamin tablets, including many multivitamin preparations plus individual vitamin tablets and gel capsules.

Experiments in this section include classification of vitamins on the basis of solubility, assays of vitamins, study of the actions of coenzymes, and determination of the vitamin C content in medicinal and commercial products.

## EXPERIMENT 29: VITAMIN CATEGORIES AND REACTIONS

Vitamins are generally divided into two major groups: fat-soluble and water-soluble. The fat-soluble ones, which are usually associated with the lipids of natural foods, include vitamins A, D, E, and K. The vitamins of the B complex and vitamin C comprise the water-soluble group.

Vitamin A is an alcohol with a high molecular weight known as retinol and is found only in the animal kingdom, occurring mainly as an ester with higher fatty acids in the liver, kidneys, lungs, and fat deposits. Vitamin A in plants occurs as carotene or carotenoid pigments.

Retinol

β-Carotene

The D vitamins are actually a group of compounds, known as sterols, that occur in nature chiefly in animals. They include ergosterol (vitamin $D_2$) and 7-dchydrocholesterol (vitamin $D_3$). The structures for these are as follows:

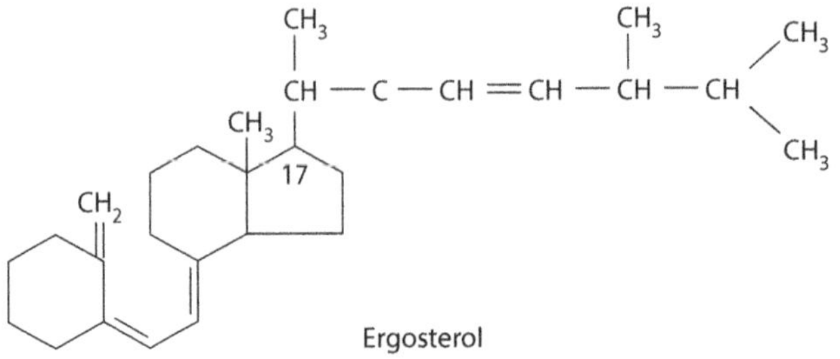

Ergosterol

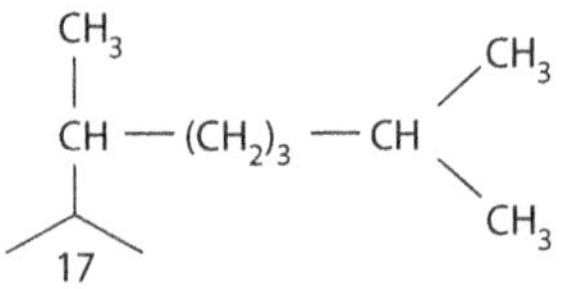

7-dehydrocholesterol

same as ergosterol except that the side chain in position 17 is that of cholesterol.

The E vitamins are known chemically as tocopherols, which are designated as α, β, and γ tocopherol Their most striking chemical characteristic is their antioxidant property. Wheat germ oil is particular rich in the E vitamins, as are milk, eggs, cereals, and leafy vegetables. The structure of α-tocopherol is:

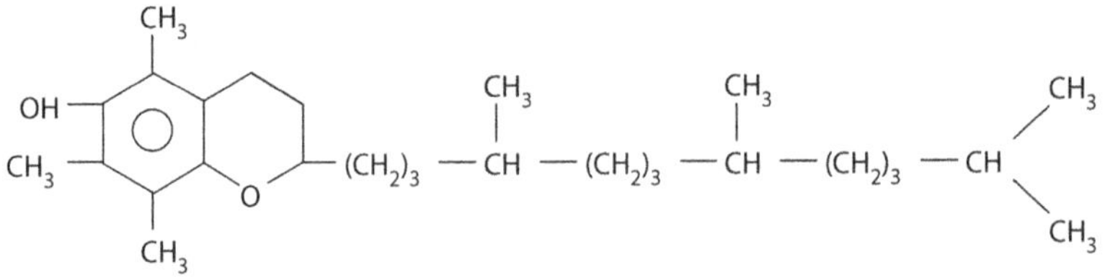

The K vitamins are related to 2-methyl 1,4 naphthoquinone and include phytonadione (vitamin $K_1$), and menadione (vitamin $K_3$). They are required for synthesis of prothrombin in the blood, which promotes proper clotting as well as being an essential component of the phosphorylation process involved in photosynthesis. Their structures are

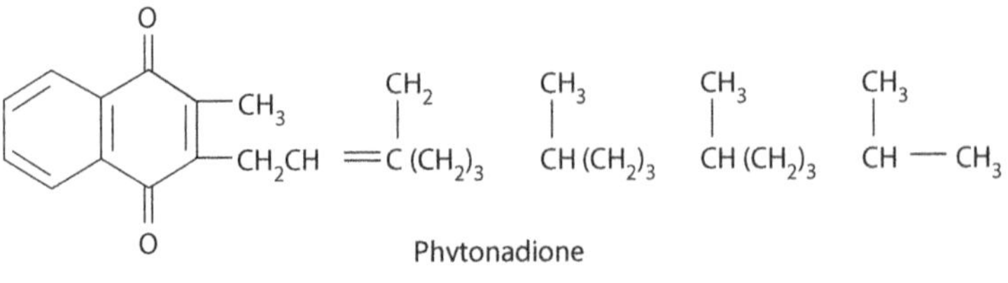

Phvtonadione

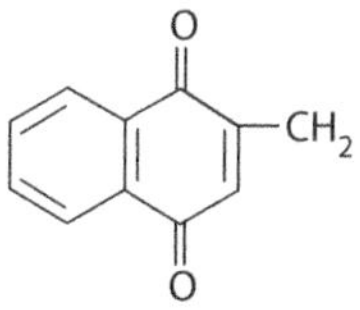

Menadione

The chemical structure of vitamin C (ascorbic acid) resembles that of a monosaccharide. Ascorbic acid functions as a reducing agent in both plant and animal tissues. Its structure is

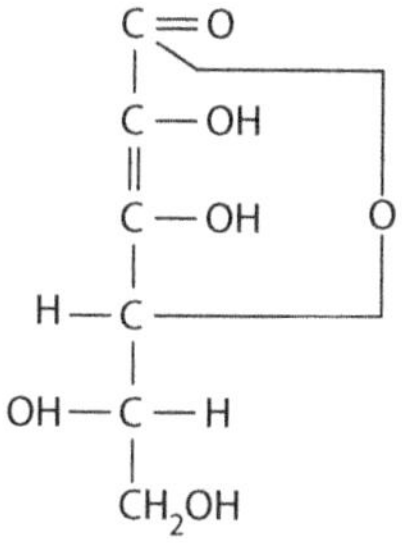

The B vitamin complex includes the following:

| | |
|---|---|
| Thiamine—B$_1$ | Pantothenic acid |
| Riboflavin—B$_2$ | Lipoic acid |
| Niacin—B$_3$ | Biotin |
| Pyridoxine—B$_6$ | Para-aminobenzoic acid (PABA) |
| Folic acid—B$_9$ | Inositol |
| Cyanocobalamin—B$_{12}$ | |

The structures of the B vitamin complex are as follows:

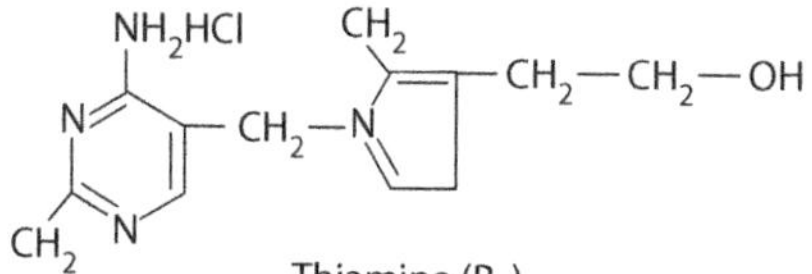

Thiamine (B$_1$)

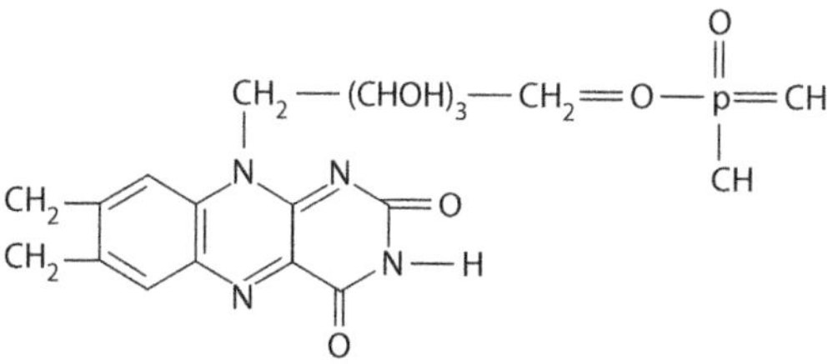

Riboflavin (B$_2$)

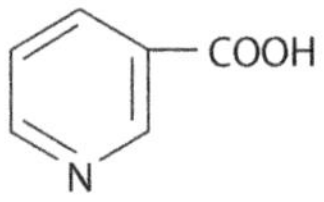

Niacin (B$_3$)

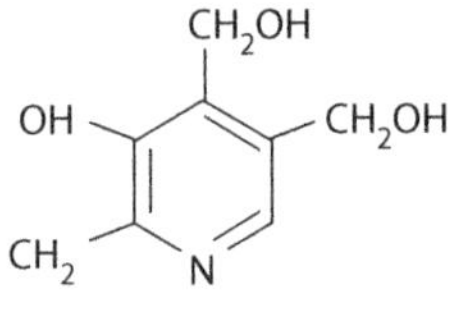

Pyridoxine (B$_6$)

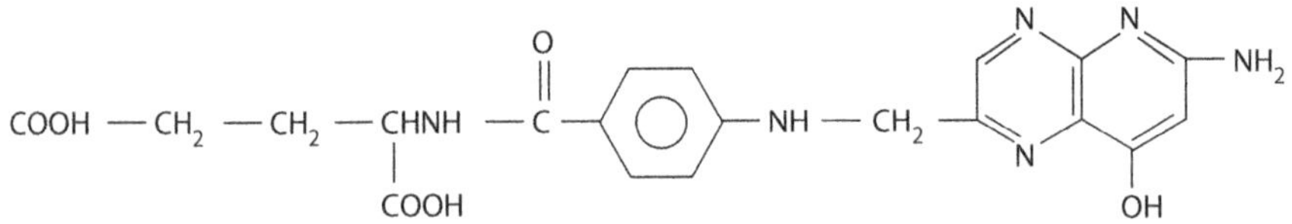

Folic acid (B$_9$)

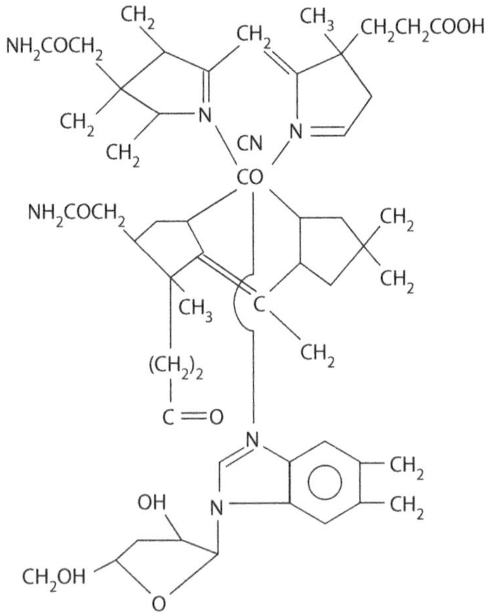

Cyanocobalamin (B$_{12}$)

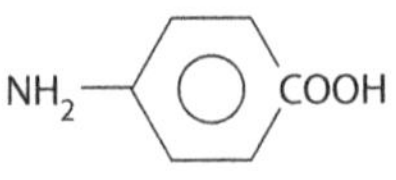

PABA

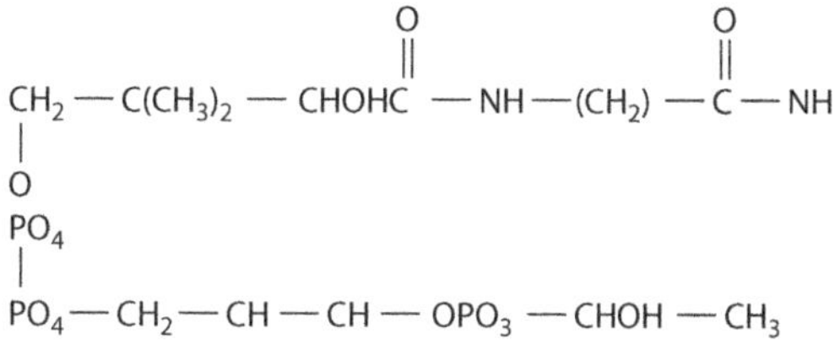

$$CH_2 - C(CH_3)_2 - CHOHC \overset{\overset{\displaystyle O}{\|}}{} - NH - (CH_2) - \overset{\overset{\displaystyle O}{\|}}{C} - NH$$

$$\underset{\displaystyle PO_4}{\overset{\displaystyle O}{|}}$$

$$PO_4 - CH_2 - CH - CH - OPO_3 --- CHOH - CH_3$$

Panthothenic acid

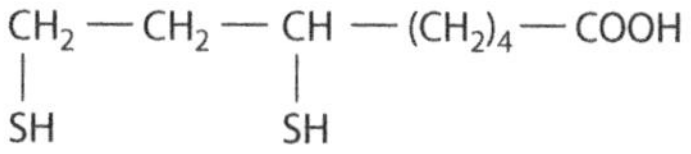

$$CH_2 - CH_2 - CH - (CH_2)_4 - COOH$$
$$\underset{\displaystyle SH}{|} \qquad \underset{\displaystyle SH}{|}$$

Lipoic acid

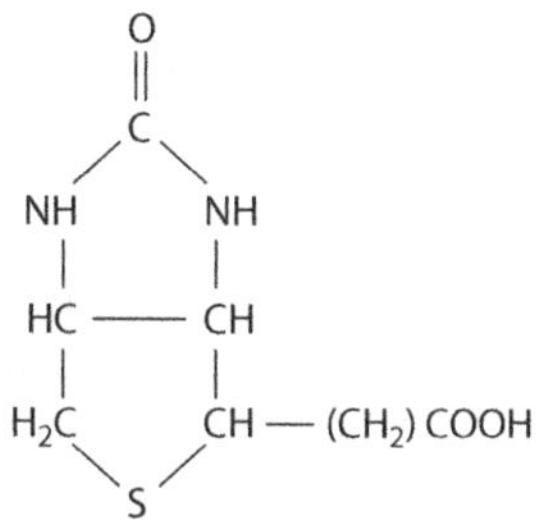

Biotin

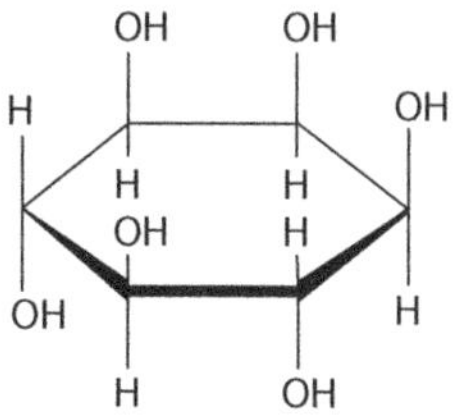

Inositol

Determining categories of vitamins is based on solubilities. In this experiment, you will make up a row of 12 test tubes, the first six of will contain 10 mL of water and the second six 10 mL of mineral oil. Number them 1 through 12. Into tubes 1 and 7 place 10 mg of riboflavin and shake each tube thoroughly. Record solubility results in the table. Repeat this procedure with the remaining vitamins listed and record results. Do your overall results confirm the traditional categories?

| Vitamin | Soluble (check ✓) | Insoluble |
|---------|-------------------|-----------|
| 1. Riboflavin | | |
| 2. Ascorbic Acid | | |
| 3. Ergosterol | | |

*Continued*

| Vitamin | Soluble (check ✓) | Insoluble |
|---|---|---|
| 4. α-Tocopherol | | |
| 5. β-Carotene | | |
| 6. Thiamine | | |
| 7. Riboflavin | | |
| 8. Ascorbic acid | | |
| 9. Ergosterol | | |
| 10. α-Tocopherol | | |
| 11. β-Carotene | | |
| 12. Thiamine | | |

# EXPERIMENT 30: THIAMINE (B₁) AS A COENZYME

Thiamine is the first member of the vitamin B complex. It acts as a coenzyme, which means it must bond to an enzyme before the enzyme is activated. The enzyme also binds the substrate. The coenzyme reacts with substrate while they are both bound to the enzyme (which is a large protein molecule). Without the coenzyme thiamine, no chemical react ion would occur. The special name given to coenzymes that are essential to the nutrition of an organism is *vitamin*. In this experiment, a benzoin condensation of benzaldehyde is carried out using coenzyme thiamine hydrochloride as the catalyst:

(benzaldehyde)                                          (benzoin)

*Procedure*: Add 1.5 g of thiamine HCl to a dry 50 mL flask and dissolve the solid in 5 mL of water by swirling. Add 15 mL of $C_2H_5OH$ and cool the solution for a few minutes in an ice bath. Place a magnet in the flask, put it on a magnetic stirrer and add 5 mL of 2 M NaOH. Now weigh the flask and solution. Then add 9 mL of benzaldehyde and reweigh to determine an accurate weight of the benzaldehyde used. Attach an air condenser and heat the reaction mixture in a water bath at 60°C for 1½ hours.

At the end of the reaction time, allow the mixture to cool to room temperature and then induce crystallization of the benzoin by cooling the mixture in ice water. If the product separates as an oil, reheat the mixture until it is once again homogenous and allow it to cool more slowly than before. When crystallization is complete, cool the mixture in an ice bath. Collect the product by vacuum filtration using a Büchner funnel and appropriate filter paper. Weigh the product and record all calculations.

Results:

Weight of thiamine and benzaldehyde _____ g.
Weight of thiamine solution only (subtract)_____ g.
Weight of benzaldehyde _____ g.

Calculate the percentage yield:

| Benzaldehyde | Benzoin |
|---|---|
| (106 m.w) | (212 m.w) |

$$\frac{106}{g} = \frac{212}{xg} \times = \text{-----} g \text{ (theoretical yield)}$$

$$\text{Percentage yield} = \frac{\text{Actual yield}}{\text{Theoretical yield}} \times 100 = \text{-----} \%$$

## EXPERIMENT 31: THIAMINE ASSAY OF VITAMIN B COMPLEX TABLETS

The object of this experiment is to determine the amount of thiamine as an ingredient of a multiple vitamin B complex tablet. It involves several very precise steps.

Begin by preparing a potassium ferricyanide ($K_3Fe(CN)_6$) solution by dissolving 1 g of the substance in 100 mL of water. Now mix 4 mL of the $K_3Fe(CN)_6$ solution with sufficient 3.5 N NaOH to make l00 mL of your oxidizing reagent.

Next, prepare a thiamine HCl stock solution by transferring 25 mg of thiamine HCl to a 1000 mL volumetric flask. Dissolve this in 300 mL of 20% $C_2H_5OH$ and adjust with 3 N Hl to a pH of 4.0.

To make the assay preparation, place in a suitable volumetric flask sufficient powder from the tablet to be assayed, such that when diluted to volume with 0.2 N HCl, the resulting solution will contain about 100 µg of thiamine HCl per milliliter. Now dilute 5 mL of this solution quantitatively and stepwise using 0.2 N HCl to an estimated concentration of 0.2 µg of thiamine HCl per milliliter.

The standard preparation is now made by diluting a portion of the stock solution, quantitatively and stepwise, with 0.2 N HCl to obtain a preparation of 0.2 µg of thiamine HCl.

The next step is to take three test tubes of about 40 mL capacity and pipet 5 mL of the standard preparation while quickly adding 3 mL of the oxidizing reagent and 20 mL of isobutyl alcohol, mixing by vigorous shaking. Now prepare a blank of the standard preparation by substituting for the oxidizing reagent an equal volume of the 3.5 N NaOH and proceeding in the same manner.

Repeat this procedure on the standard preparation with the assay preparation, including the making of a blank.

Next, into each of the eight test tubes, pipet 2 mL of 100% $C_2H_5OH$ and allow the phases to separate. Decant to draw off 10 mL of the clear supernatant isobutyl alcohol solution into standardized cells. Then measure the fluorescence in a suitable fluorometer having an input filter of narrow transmittance range with a maximum of 365 nm and an output filter with a maximum of about 465 nm.

Calculate the microgram of thiamine HCl in each 5 mL of the assay preparation using the formula:

$$(A - b)/(S - d)$$

in which A and S are the average fluorometer readings of the portions of the assay and standard preparations treated with the oxidizing reagent, and band d are the readings of the blanks of the assay preparations and the standard preparations, respectively.

Test results:

$$(A - b)/(S - d)$$

## EXPERIMENT 32: DETERMINING VITAMIN C CONTENT OF COMMERCIAL TABLETS

Vitamin C, known chemically as ascorbic acid, occurs in citrus fruits and in many vegetables as well. Its structural formula is

The purpose of this investigation will be to determine the actual amount of vitamin C present in commercial tablets advertised as having 100 mg of the substance.

Rinse and fill a 100 mL buret with 0.01 M $KIO_3$ solution and record the initial buret reading. Next weigh a 100 mg vitamin C tablet (remember, it will weigh more due to fillers and binders) and record. Put about 50 mL of deionized water in a 250 mL Erlenmeyer flask and crush and dissolve the tablet in it. Add 1 g of KI and swirl to dissolve, then add 5 mL of 1 M HCl. Now add 3 mL of a 0.5% starch indicator and titrate with the $KIO_3$ in the buret until a deep blue develops permanently. Record the final buret reading. On the basis of the molecular weight of ascorbic acid (176.13), calculate the actual amount of vitamin C in the tablet with that on the manufacturer's label. Obtain an average class value on this.

The rationale for this reaction is that the Ki (using HCl to supply the hydrogen ion) and starch solution added to the vitamin C react with the $IO_3$ ion causing the reaction:

$$IO_3^- = 5I^- = 6H = 3I_2 = 3H_2O$$

Then, almost as quickly as it is formed, the ascorbic acid reacts with the $I_2$. When all the ascorbic acid has reacted, the concentration of the $I_2$ begins to build to a level where it will react with $I^-$ to form the linear triiodide

$$I_2 = I^- \; I_3^-$$

Triiodide ion combines with the starch to form the starch–triiodide ion complex that appears as a deep blue-black color signaling the end point of the titration.

Results:

Weight of (100 mg) Vitamin C tablet _____
Volume of $KIO_3$—initial reading _____
Volume of $KIO_3$—final reading _____
Volume of $KIO_3$ required to titrate _____

Calculate the mass of the vitamin C (176.13 g/mol).

## EXPERIMENT 33: COLOR TESTS OF SOME FAT-SOLUBLE VITAMINS

Retinol (vitamin A): Prepare a 6 µg solution of retinol in chloroform (CHCl3) by dissolving 0.006 g of retinol in 1 liter of the $CHCl_3$. Pipet 1 mL of this solution into a small test tube and add 10 mL of $SbCl_3$ T.S. (if none in stock, you can make some by dissolving 20 g of $SbCl_3$, in sufficient $CHCl_3$ to make 100 mL).

Resulting color: _____.

Vitamin $D_2$ and $D_3$ tablets come as a mixture of ergocalciferol and cholecalciferol, which give identical color reactions. Place 0.5 mg of the tablet in 5 mL of $CHCl_3$, then add 0.3 mL of acetic anhydride and 0.1 mL of concentrated $H_2SO_4$. Shake.

Resulting color changes: _____.
(Note: the initial color is not the final one on this.)

Vitamin E occurs as α, β, and γ tocopherols and occur together in soft gel capsules sold over-the-counter in pharmacies. The characteristic color reaction of the tocopherols can be achieved by taking 10 mL of the vitamin E and adding 2 mL of concentrated $HNO_3$. Heat at 75° for 15 minutes.

Resulting color: _____.

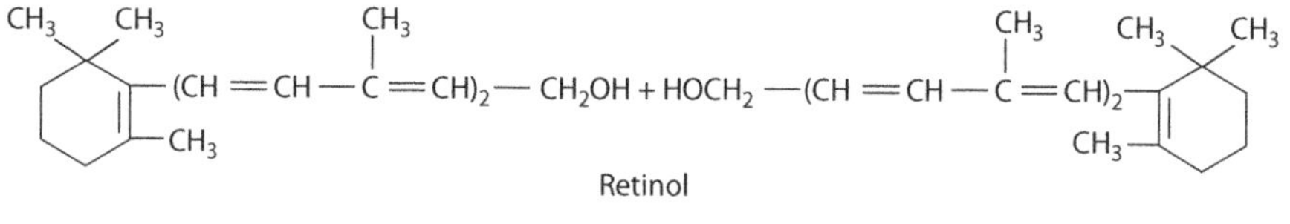

Retinol

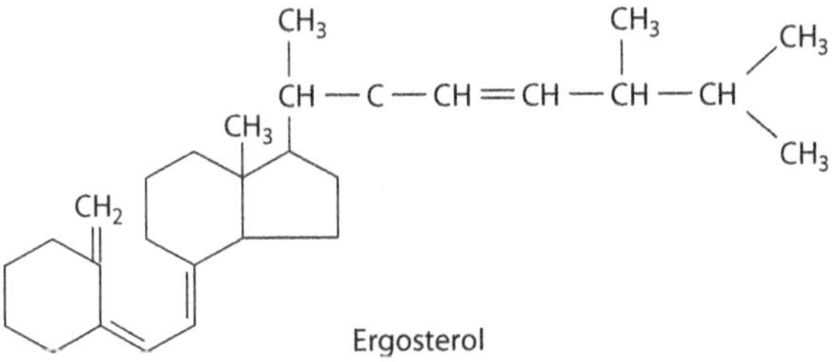

Ergosterol

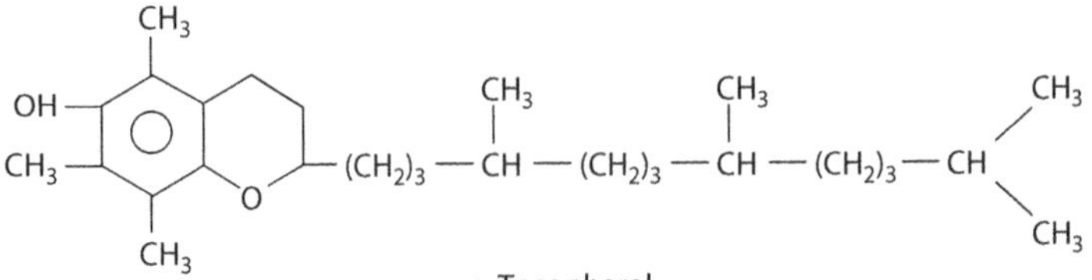

α-Tocopherol

## SELECTED REFERENCES—VITAMINS

Augustin, J. *Methods of Vitamin Analysis*. Wiley, 1985.
Ball, G. *Vitamins in Food*. CRC Press, 2005.
Bender, D. *Nutritional Biochemistry*. Cambridge University Press, 1992.
Coward, K. *Biological Standardization*. Wilkins, 1947.
Johnson, H. C. *Methods of Vitamin Determination*. Burgess, 1948.
Kirk, R. *Encyclopedia of Chemical Technology*. Wiley, 1997.
Moulton, G. *Chemistry and Biology of Vitamins*. Penguin, 2004.
Munson, P. *Vitamins and Hormones*. Academic Press, 1981.
Nelson, D. *Principles of Biochemistry*. Freeman, 2008.
Rosenberg, H. *Chemistry and Physiology of Vitamins*. Interscience, 1945.
Rucker, R. *Handbook of Vitamins*. CRC Press, 2007.
Sebrell, W. H. *The Vitamins* (3 volumes). Academic, 1954.

# 7 Steroids

## STEROIDS

Steroids are among the most useful drugs in modern medicine. While they are united by the presence of a cyclopentanoperhydrophenanthrene ring system, their therapeutic effects are quite diverse. From a physiological perspective, they may act as cardiac agents or diuretics, antibiotics, sex hormones or as digestants, vitamin precursors, neuromuscular blockers, contraceptives, and anti-inflammatory agents.

Some commonly prescribed steroids in modern medicine include spironolactone used as a diuretic, prednisone used as an anti-inflammatory, norethindrone used as a female contraceptive, and digitoxigenin, employed as a cardiac drug.

Experiments in this section focus on color reactions of specific steroids such as hydrocortisone, cholesterol, and estradiol.

## EXPERIMENT 34: COLOR REACTIONS OF VARIOUS STEROIDS

Steroids comprise a natural product widely distributed throughout the plant and animal world. Although they have widely variant pharmacological actions, they are defined as compounds having a cyclopentanoperhydrophenanthrene nucleus:

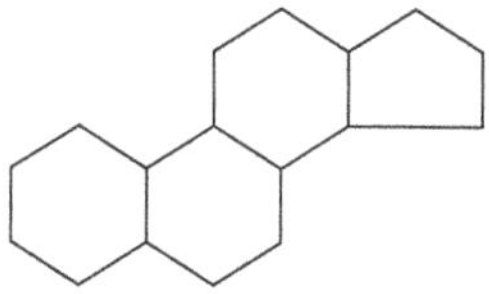

Steroids are included in such compounds as bile acids, cardiac glycosides, steroid hormones, and vitamins. Some representative structures are shown here.

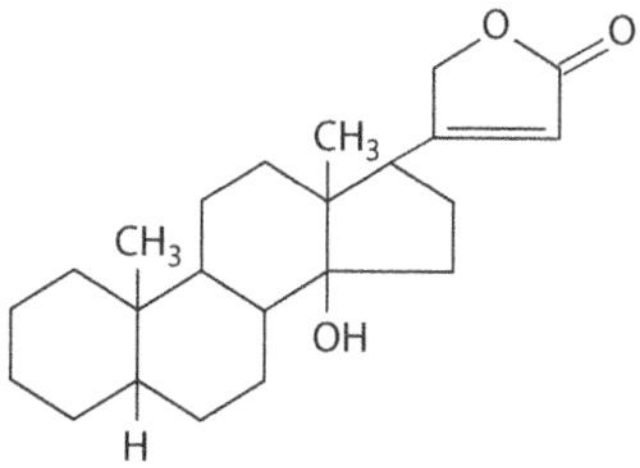

Cholesterol

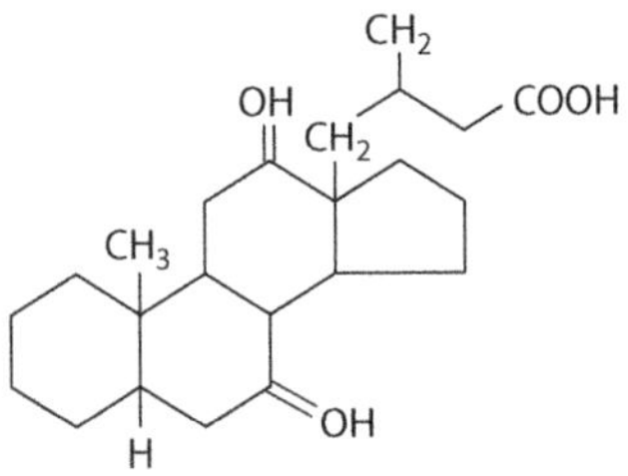

Cholic Acid

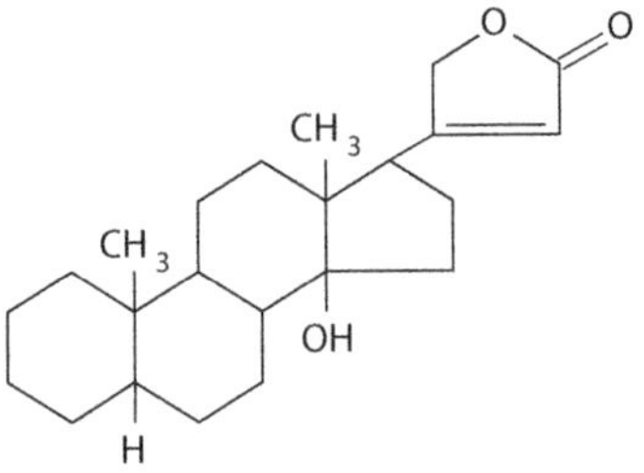

Digitoxigen

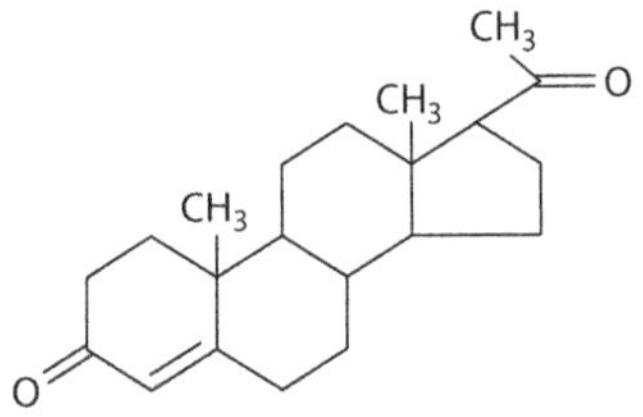

Progesterone

Hydrocortisone is an example of a steroid that exhibits distinct color reactions to the treatment of reagents. Its structure is:

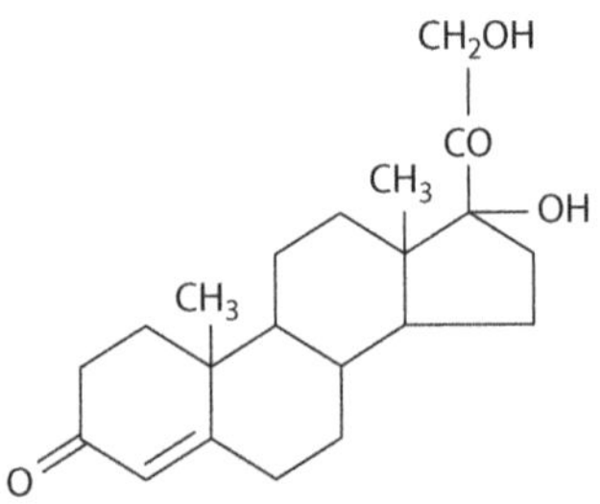

Dissolve 2 mg of hydrocortisone powder in 2 mL of $H_2SO_4$ and allow it to stand for 15 minutes.

Record color reaction: _____.

Now dilute the solution with 10 mL of water and record both color and other reactions.

Results: _____.

Digitoxin is a cardiac glycoside obtained from the digitalis purpurea (purple foxglove) plant. Begin preparing an assay mixture by dissolving 20 mg of digitoxin powder in 20 mL of chloroform (CHCl$_3$). Transfer this solution to a 100 mL volumetric flask and bring it to volume with benzene (C$_6$H$_6$). Next evaporate the 100 mL on a steam bath to dryness. Dissolve the residue in 2 mL of a solution prepared by mixing 0.3 mL of FeCl$_3$ t.s. and 50 mL of glacial acetic acid and underlay with 2 mL of H$_2$SO$_4$. Record the sequence of color reactions.

Results: _____.

Cholesterol tests. To a solution of 10 mg of cholesterol in 1 mL of chloroform (CHCl$_3$) add 1 mL of H$_2$SO$_4$.

Results: _____.

Now dissolve 5 mg of cholesterol in 2 mL of CHCl$_3$ and add 1 mL of acetic anhydride. Record sequences of colors.

Results: _____.

The final tests of this section are with the hormone estradiol benzoate whose structure is

Dissolve 2 mg of estradiol benzoate in 2 mL of H$_2$SO$_4$.

Results: _____.

Next dissolve 100 mg of estradiol benzoate in 10 mL of CH$_3$OH and add 100 mg of K$_2$CO$_3$ dissolved in 0.5 mL of water; reflux the mixture on a steam bath for 2 hours. Add 30 mL of water and heat gently until all the alcohol is evaporated. Then add 15 mL of water and keep the solution at a temperature of between 5 and 10°C for 1 hour. Filter the precipitate, wash it with cold water until the washings are neutral to litmus paper, and dry at 100°C for 1 hour.

Set up a melting point apparatus as shown and determine the melting point of the estradiol benzoate.

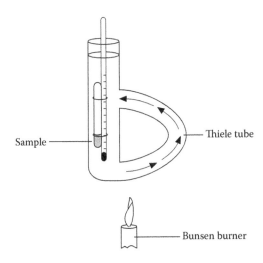

Melting point determined: _____.

Check the USP/NF to determine the accuracy of your measurement.

## SELECTED REFERENCES—STEROIDS

Bodem, G. *Cardiac Glycosides*. Springer, 1978.
Butt, W. *Hormone Chemistry*. Horwood, 1976.
Dence, J. B. *Steroids and Peptides*. Wiley, 1980.
Fried, J. *Organic Reactions in Steroid Chemistry*. Yoyoman, 1972.
Gower, D. W. *Steroid Hormones Yearbook*. Medical, 1975.
Lednicer, D. *Steroid Chemistry at a Glance*. Wiley, 2010.
Makin, H. L. *Steroid Analysis*. Springer, 2010.
Pavia, D. *Organic Laboratory Techniques*. Saunders, 1988.
Witzman, R. F. *Steroids: Keys to Life*. Van Nostrand, 1981.

# 8 Plant Acids

## PLANT ACIDS

While plant acid products do not comprise a specific pharmaceutical group as such they are nevertheless extremely important in everyday use. One of the most widely used is nicotinic acid (niacin), which is part of the Vitamin B complex.

Another plant acid is tannic acid, which has significant astringent properties and has, therefore, been employed as an ointment for the treatment of burns. Salicylic acid, from the willow, is a component of the widely used anti-inflammatory, antipyretic chemical known as acetylsalicylic acid (aspirin).

Other significant plant acids include oxalic in rhubarb, acetic in vinegar, cinnamic in cinnamon, and citric in various citrus fruits.

The experiments in this section deal with isolating citric acid from lemon, plus using a titration procedure to assess the molarity of acetic acid in vinegar.

## EXPERIMENT 35: ISOLATION OF CITRIC ACID FROM LEMON

Plant acids are defined as simple organic compounds containing not more than six carbon atoms and two or three carboxyl groups. Many of them function as intermediates in cellular respiration. Several, such as citric acid, malic acid, and succinic acid, occur widely in plant tissues in rather high concentrations. Others, less well known, that occur in smaller quantities, include α-ketoglutaric acid and cis-aconitic acid.

Citric acid is one of the most widely distributed plant acids, occurring in cranberries, red currants, strawberries, raspberries, beets, citrus fruit, and animal tissues. It may be isolated from citrus fruits as the comparatively insoluble calcium salt. This compound displays unusual solubility properties in that it becomes less soluble at increased temperatures. This information should be kept in mind when carrying out the following experiment. The formula for citric acid is:

$$OH- \underset{\underset{\text{CH}_2-\text{COOH}}{|}}{\overset{\overset{\text{CH}_2-\text{COOH}}{|}}{C}} -COOH$$

Measure 90 mL of thawed frozen lemon juice concentrate into a 250 mL beaker and carefully add a 10% NaOH solution with stirring until the mixture is slightly alkaline. A distinct color change occurs at this point, the solution passing from a clear yellow to a brownish color. Strain the solution through muslin to remove large particles of pulp and then filler through paper in a Büchner funnel. The pores of the filter paper may tend to become clogged by the extract in spite of the previous straining. Should this occur, change the paper in the funnel once or twice as required to complete the filtration.

Measure the filtrate and record.

Result: _____ mL.

Now place the filtrate in a beaker and add 5 mL of 10% $CaCl_2$, stirring constantly for each 10 mL of filtrate. Heat to boiling and filter off the copious precipitate of calcium citrate ($Ca_3C_{12}H_{10}O_{14}$) from the hot solution using a Büchner funnel. Wash the precipitate with a small quantity of boiling water. Now resuspend it in a minimum quantity of cold water, heat to boiling, and once more collect the insoluble $Ca_3C_{12}H_{10}O_{14}$ by filtration. Allow the salt to air dry. Wash and calculate the yield.
Calculations:
Citric acid may be prepared from the citrate salt by weighing the air-dried salt and placing it in a beaker. Add sufficient 1 N $H_2SO_4$ required to convert the salt to the acid. The equation for the reaction is:

$$Ca_3C_{12}H_{10}O_{14} + 3\ H_2SO_4 \rightarrow C_{12}H_{10}O_{14} + 3\ CaSO_4$$

Allow the mixture to stand for a few minutes, filter off the insoluble $CaSO_4$, and concentrate the filtrate to a small volume in a steam bath. Citric acid crystallizes out. Filter, dry, and weigh the acid. Calculate the percentage of citric acid in the lemon juice sample you used.
Calculations:

## EXPERIMENT 36: MOLARITY OF ACETIC ACID IN VINEGAR

Vinegar occurs as the product of acetic fermentation of alcoholic products, usually wine, cider, or malt. The active principle of vinegar is acetic acid, which varies in amount from 2% to 15%. The formula for acetic acid is

$$CH_3COOH$$

The amount of acetic acid in a sample of white vinegar will be determined in this experiment by titrating the vinegar with a solution of NaOH whose concentration is known. The indicator used is phenolphthalein.

Begin by placing 30 mL of white vinegar in a clean, dry 150 mL beaker. Then place 85 mL of a 0.3 M NaOH solution in an Erlenmeyer flask and close immediately with a rubber stopper.

Fill a 100 mL buret with the NaOH solution and pipet 10 mL of the vinegar into a 125 mL Erlenmeyer flask. Add 2 drops of the phenolphthalein indicator to the acid water mixture.

Record the molarity of the NaOH solution and the initial buret reading. Place the flask containing the vinegar under the buret and begin titration in increments of 1 mL while swirling in a magnetic stirrer. The titration is complete when the addition of about 1 mL of NaOH causes the color in the flask to change from colorless to a shade of pink. Record the buret reading at this point and subtract the initial reading from this reading to ascertain the volume required for the approximate end point.

You should repeat this procedure two additional times to obtain an average or mean figure for the three titrations.

Results:

| Sample Number | 1 | 2 | 3 |
|---|---|---|---|
| Final buret reading | _____ | _____ | _____ |
| Initial reading | _____ | _____ | _____ |
| Volume of NaOH | _____ | _____ | _____ |
| | Average of above 3 | | _____ |

Molarity formula $M_1V_1 = M_2V_2$.
($V_2$ = acid volume, which is 10 mL; $M_1$ = base molarity, which is 0.3; and $V_1$ = average volume of base used.)

## SELECTED REFERENCES—PLANT ACIDS

Bonner, J. *Plant Biochemistry*. Academic Press, 1976.
Daniel, M. *Medicinal Plants*. CHIPS, 2006.
Gleason. F. *Plant Biochemistry*. Jones & Bartlett, 2011.
Gokhale, S. B. *Practical Pharmacognosy*. Prakashan, 2008.
Goodwin, T. *Chemistry and Biology of Plant Pigments*. Academic Press, 1965.
Helot, H. W. *Plant Biochemistry*. Elsevier, 2010.
Rosenthaler, L. *The Chemical Investigation of Plants*. Bell, 1930.
Shelland, H. J. *Practical Plant Chemistry*. Pitman, 1957.
Tyler, V. *Experimental Pharmacognosy*. Burgess, 1962.
Wentworth, R. *Experiments in General Chemistry*. Houghton, 1993.

# 9 Flavonoids

## FLAVONOIDS

Flavonoids have been reported by researchers to exist in 137 different natural states, distributed through at least 62 botanical families, 153 genera, and 277 species. They have been employed for a variety of purposes, ranging from the ability to reduce capillary fragility to protection against acute toxicity and from exhibiting estrogenic activity to counteracting inflammation in a manner similar to that of cortisone.

Chemically the flavonoids are usually glycosides with phenolic hydroxyl groups.

Modern medicine makes use of several flavonoids, including hesperetin, used in combination with ascorbic acid to reduce capillary fragility in cerebrovascular and cardiovascular diseases, hypertension, and hemorrhagic nephritis. Another flavonoid product is quercetin sold in health food stores. Quercetin is a flavonoid glycoside sold as a possible antidote for high cholesterol.

Experiments in this section include the isolation of hesperidin from orange peel and naringin from grapefruit peel, plus a chromatography experiment on various flavonoids.

## EXPERIMENT 37: ISOLATION OF HESPERIDIN FROM ORANGE PEEL

Flavonoid compounds usually occur in plants as glycosides in which one or more of the phenolic hydroxyl groups are combined with sugar residues. Flavonoids occur in all parts of plants, including the fruit, pollen, roots, and hardwood. Numerous physiological activities have been attributed to them; thus, small quantities may act as cardiac stimulants. Others like hesperidin appear to strengthen weak capillary blood vessels, while still others act as diuretics and antioxidants for fats.

The molecular structure of hesperidin, which is the major flavonoid present in lemons and sweet oranges:

Hesperidin was first isolated in 1828 from the spongy inner portion of the peel of oranges and later from lemons. There are several effective methods. The one that we will use consists of extracting the dry peel successively with petroleum ether and then with methanol, the first solvent removing the essential oil and the second, the glycoside.

*Procedure for Extracting Hesperidin:* Place 200 g of dried orange peel in a round-bottomed flask attached to a reflux condenser. Add 1 liter of petroleum ether and heat on a water bath for 1 hour. Filter the contents of the flask through a Büchner funnel and allow the powder to dry at room temperature. Now return the dry powder to a flask and add 1 liter of methanol and heat under reflux for 3 hours. Filter this hot using a vacuum. This leaves a syrupy residue, which may then be crystallized from dilute acetic acid yielding white needles with a melting point of 252°C–254°C.

*Purification of Hesperidin:* To a 10% solution of the crude hesperidin in formamide prepared by warming to 60°C and treating with activated charcoal that has been boiled with dilute HCl, filter through celite diluted with an equal volume of water and allow to stand for a few hours to crystallize. The crystals of the purified hesperidin are then filtered off and washed first with hot water and then with isopropanol yielding a white crystalline product.

Record results:

Melting point of the hesperidin _____ °C
Weight of the hesperidin _____ g.

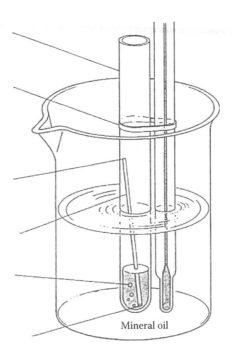

Mineral oil

## EXPERIMENT 38: ISOLATION OF NARINGIN FROM GRAPEFRUIT PEEL

The spongy inner portion of grapefruit peel is composed chiefly of cellulose, soluble carbohydrates, pectins, amino acids, vitamins, and flavonoids. The flavonoid is called naringin, which has an intensely bitter taste and a molecular formula of

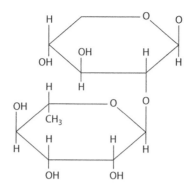

*Procedure for Isolation*: Place 100 g of grapefruit peel and 400 mL of water and heat in a boiling flask at 90°C for 5 minutes. The water extract is then filtered off and saved. Add 50 mL water to the solid and heat to 80°C for 5 minutes and filter. Next, combine the two extracts and boil with 1% cellite, filter, and concentrate in vacuo to approximately one-ninth the original volume. Crystallize the concentrated extract in a refrigerator and then filter.

Next, dissolve 8.6 g of naringin in 100 mL of boiling isopropanol and filter hot. Now heat the filtrate to its boiling point to initiate crystallization. Allow it to cool and filter on a Büchner funnel wash with cold isopropanol.

Record results:

Melting point of the naringin _____°C
Weight of the naringin _____ g.

## EXPERIMENT 39: PAPER CHROMATOGRAPHY OF FLAVONOIDS

Paper chromatography has been applied to flavonoid pigments for more than half a century. The reason for extensive use of the method is that the natural color of most of these compounds in visible light permits the ready distinction of the anthocyanins and deeply colored aurone and chalcone pigments and ultraviolet light reveals most of the other flavonoids. On spraying, the flavonoid compounds may be converted into more deeply colored or fluorescent derivatives.

*Procedure for Chromatography*: Cut strips of Whatman No. 1 filter paper into 2.5 × 50 cm and spot them with the following pigments in alcoholic solutions:kaempferol, morin, quercetin, rhamnetin, quercetin, rutin, apigenin, chrysin, hesperidin, naringin, and neohesperidin.

Use a hair dryer to evaporate the spotting solvent. Then place the strips in a chromatographic chamber with the ends beneath the surface of a solvent system composed of 50:50 ethyl acetate and water.

Record results using three solvent systems:

1. Ethyl acetate–water 50:50
2. Phenol–water 50:50
3. Isopropanol–water 60:40

Compare your results with the following table

| Flavonoid | R_f System 1 | 2 | 3 |
|---|---|---|---|
| Kaempferol | 0.90 | 0.74 | 0.77 |
| Morin | 0.71 | 0.65 | 0.58 |
| Quercetin | 0.81 | 0.42 | 0.67 |
| Rhamnetin | 0.92 | 0.71 | 0.73 |
| Quercitin | 0.50 | 0.56 | 0.79 |
| Rutin | 0.15 | 0.45 | 0.83 |
| Apigenin | 0.87 | 0.89 | 0.89 |
| Chrysin | 0.86 | 0.93 | — |
| Hesperidin | 0.77 | 0.85 | 0.63 |
| Naringin | 0.51 | 0.75 | 0.86 |
| Neohesperidin | 0.38 | 0.75 | 0.88 |

Report your results:

## SELECTED REFERENCES—FLAVONOIDS

Andersen, O. *Flavonoid Chemistry.* CRC Press, 2005.
Andersen, O. *Flavonoids: Chemistry, Biochemistry, and Applications.* CRC Press, 2006.
Bohn, B. *Introduction to Flavonoids.* Medical Publ., 1988.
Geissman, T. *The Chemistry or Flavonoids.* Pergamon, 1962.
Grotewold, E. *The Science of Flavonoids.* Springer, 2007.
Harborne, J. B. *Biochemistry or Phenolic Compounds.* Academic Press, 1965.
Keller, R. *Flavonoids: Biosynthesis, Biological Effects, and Dietary Sources.* Nova Science, 2009.
Sinclair, W. B. *The Citrus Flavonoids.* University of California Press, 1961.
Vernin, G. *The Chemistry of Heterocyclic Flavonoids.* Horwood, 1982.

# 10 Alkaloids

## ALKALOIDS

A precise definition of alkaloids is difficult to develop since they do not represent a homogeneous group of compounds. Except for the fact that they are all organic nitrogenous compounds and possess basic properties, it is difficult to arrive at any general description. Alkaloids are named on the basic of

1. the generic name of the plant yielding them, or
2. the specific name of the plant yielding them, or
3. the common name of the drug yielding them, or
4. their physiological activity, or occasionally
5. the name of the discoverer.

Alkaloids find extensive use in modern medicine such as belladonna used as a sedative, strammonium as a parasympatholytic, quinine as an antimalarial, ipecac as an emetic, cocaine as an anesthetic, and opium derivatives used in the production of morphine and codeine.

The experiments in this section include color detection and chromatography of alkaloids, the isolation of some alkaloids from various plant sources, which include strychnine from nux vomica, caffeine from tea, and berberine from goldenseal.

## EXPERIMENT 40: COLOR SPOT TESTS FOR DETECTION OF ALKALOIDS

### INTRODUCTION

The earliest recorded studies on vegetable alkalies (later termed *alkaloids*) date back to 1817 with the German chemist, Sertürner, and his isolation of morphine. Most of the alkaloids have been discovered by pharmacists who have obviously been interested in them for their physiological activity. Some alkaloids come from the cryptogams (nonflowering plants), but most are from the phanerogams.

Some unifying properties that distinguish the alkaloids include:

1. They contain nitrogen, in addition to carbon and hydrogen.
2. They are usually nonvolatile when solid, but volatile when liquefied.
3. They unite with acids to form ammonium salts, and therefore they are precipitated by bases.
4. Most are physiologically active, some are poisonous.
5. They are mainly crystallizable, but some are amorphous.

6. Most are white in color except berberine and sanguinaria salts.
7. They exhibit optical activity.
8. Most are insoluble in water, but soluble in organic solvents.
9. Most are precipitated by Mayer's Marme's, Dragendorff's, Wagner's, Sonnenschein's, and Schreibler's reagents, as well as by gold chloride, tannic acid, and picric acid.

Some of the major alkaloidal test solutions are

Wagner's reagent—2 g of iodine and 6 g of KI in 100 mL of water
Mayer's reagent—1.358 g of $HgCl_2$ and 5 g of KI in 100 mL water
Dragendorff's reagent—potassium iodide and bismuth iodide
Marme's reagent—2 g of $CdI_2$ and 4 g of KI in 12 mL water
Sonnenschein's reagent—phosphomolybdic acid solution
Schreibler's reagent—10% solution of phosphotungstic acid
Erdmann's reagent—10 drops of $HNO_3$ in 20 mL of concentrated $H_2SO_4$
Froehde's reagent—1 g of $NH_4MoO_4$ in 100 mL of $H_2SO_4$
Mandelin's reagent—1 g of $NH_4VO_4$ in 200 g of $H_2SO_4$
Ferric chloride solution—8.2 g of $FeCl_3$ in 100 mL of water
Tannic acid solution—1 g of tannic acid in 1 mL $CH_3OH$ and 10 mL water
Picric acid solution—1 g of picric acid in 100 mL of water
0.1 N iodine T.S.—14 g of I and 36 g of KI in 1000 mL of water

*Alkaloidal classification*: The alkaloids are generally classified by their molecular structure and fall into 11 major categories: pyridine–piperidine, tropane, quinoline, isoquinoline, indole, imidazole, steroidal, lupinane, alkaloidal amines, and purine, plus a miscellaneous group.

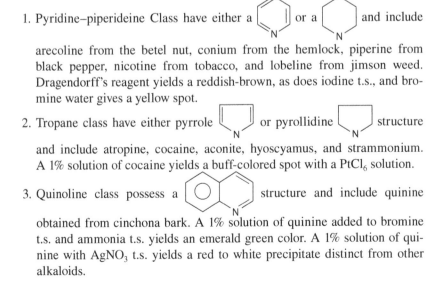

1. Pyridine–piperideine Class have either a ⬡ or a ⬡ and include arecoline from the betel nut, conium from the hemlock, piperine from black pepper, nicotine from tobacco, and lobeline from jimson weed. Dragendorff's reagent yields a reddish-brown, as does iodine t.s., and bromine water gives a yellow spot.

2. Tropane class have either pyrrole ⬠ or pyrollidine ⬠ structure and include atropine, cocaine, aconite, hyoscyamus, and strammonium. A 1% solution of cocaine yields a buff-colored spot with a $PtCl_6$ solution.

3. Quinoline class possess a ⬡⬡ structure and include quinine obtained from cinchona bark. A 1% solution of quinine added to bromine t.s. and ammonia t.s. yields an emerald green color. A 1% solution of quinine with $AgNO_3$ t.s. yields a red to white precipitate distinct from other alkaloids.

4. Isoquinoline class possess a  structure and include ipecac, hydrastis from goldenseal, sanguinaria from the bloodroot, and opium from select species of oriental poppies. A sanguinaria solution in the presence of 5 mL of $H_2SO_4$ and a drop of $FeCl_3$ yields a blue color, which upon addition of $HNO_3$ turns to dark red brown. A milligram of morphine sulfate plus 0.5 mL $H_2SO_4$ and a drop of formaldehyde yields a violet color, which changes to bluish violet. This can be used to distinguish morphine from codeine, the latter of which begins with a blue-violet color.

5. Indole class possess a  structure and include strychnine from the nux vomica plant, reserpine from the rauwolfia, plus physostigmine and ergot. When $H_2SO_4$ and a 1% solution of $NH_4VO_4$ are added to a strychnine solution the color changes on standing from violet to blue to purple to red. With 1.0 mL of $H_2SO_4$ and $K_2Cr_2O_7$ a deep blue color appears that gradually changes to violet, to cherry-red, to orange, ending in yellow.

6. Imidazole class, containing this basic structure, includes pilocarpine from the plant *Pilocarpus jaborandi*. Pilocarpine yields a red color when treated with a solution of sodium nitroprusside and sodium hydroxide, which on addition of sodium thiosulfate, turns green. With $K_4Fe(CN)_6$ a yellow color is produced which eventually turns to blue. A mixture of $NH_4VO_4$ and $H_2SO_4$ produces a sequence of colors from yellow to green and finally blue.

7. Steroidal class, which have a basic steroid nucleus,

is also known as a cyclopentenophenanthrene system. This class includes the veratrum alkaloids, as well as aconite and larkspur. A solution of aconite to which is added $CH_3COOH$ plus $KMnO_4$ yields a red crystalline precipitate, with $H_3PO_4$ a violet color is produced, and with $HVO_4$ an orange color results.

8. Lupinane class have a basic structure of which sparteine is an example. When sparteine is mixed with either $CaI_2$, phosphomolybdic acid or Mayer's reagent, a white precipitate results. With $PtCl_6$ there is a yellow precipitate, and with copper salts, a green color occurs.

9. Alkaloidal amines class have an aliphatic base plus an aromatic nucleus such as the structure of ephedrine. Colchicine also belongs to this class. Colchicine, with a couple of drops of concentrated $H_2SO_4$, produces a lemon

yellow color. When $HNO_3$ is added to this, the color sequence moves from greenish-blue to violet to red to yellow to colorless. With $FeCl_3$ t.s. an alco-

holic solution of colchicines turns garnet red.

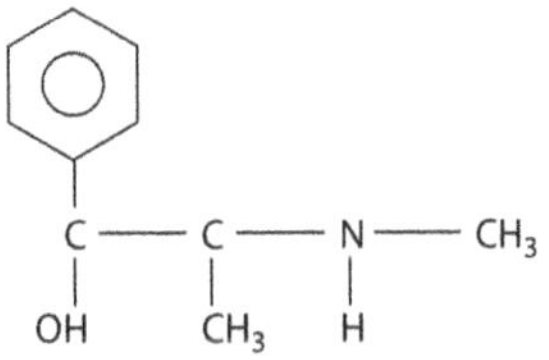

10. Purine class all begin with a purine system and include caffeine, theophylline, and theobromine. Fifty milligrams of theobromine healed with 1 mL of $AgNO_3$ solution (10%) and 6 mL of 10% NaOH solution yield a brown gelatinous mass that cannot be poured from the vessel.

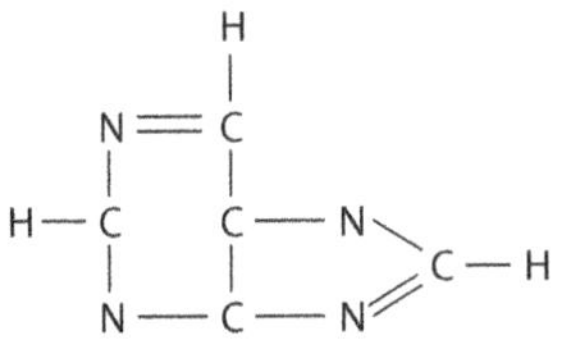

## EXPERIMENTS IN CHROMOGENIC ANALYSIS

Color tests are widely employed for various medicinal substances. Many of these tests serve to characterize a drug by the color of solution or precipitate formed when the drug is brought into contact with a suitable reagent. The color reagents function in a variety of ways. Some, such as sulfuric acid (in Froehde's and Mandelin's reagents), are powerful dehydrating agents. Others, like nitric acid, are strong oxidizers, while others rely on special reactions, such as the use of ferric chloride for phenolic alkaloids.

The following reactions are carried out by placing a few milligram of the alkaloid in the depression of a white spot plate and applying the reagent with a glass rod or dropper. It is *important* that each reaction be observed for several minutes, since the colors formed may be transient or change rapidly through several successive shades.

Line up spot plates in the following sequence. Test reagents with blanks are listed below each spot. Record reactions, including any transient color changes in each case.

| Atropine (Tropane) | | Sparteine (Lupinane) | | Quinine (Quinoline) | |
|---|---|---|---|---|---|
| $H_2SO_4$ | _____ | $H_2SO_4$ | _____ | $H_2SO_4$ | _____ |
| Erdmann's | _____ | Erdmann's | _____ | Erdmann's | _____ |
| Froehde's | _____ | Froehde's | _____ | Froehde's | _____ |
| Mandelin's | _____ | Mandelin's | _____ | Mandelin's | _____ |
| $FeCl_3$ | _____ | $FeCl_3$ | _____ | $FeCl_3$ | _____ |

*Continued*

| Strychnine (Indole) | | Pilocarpine (imidazole) | | Ephedrine (Alk. amine) | |
|---|---|---|---|---|---|
| $H_2SO_4$ | _____ | $H_2SO_4$ | _____ | $H_2SO_4$ | _____ |
| Erdmann's | _____ | Erdmann's | _____ | Erdmann's | _____ |
| Froehde's | _____ | Froehde's | _____ | Froehde's | _____ |
| Mandelin's | _____ | Mandelin's | _____ | Mandelin's | _____ |
| $FeCl_3$ | _____ | $FeCl_3$ | _____ | $FeCl_3$ | _____ |

Given the results of the spot tests conducted, suggest ways of differentiating between the following pairs of alkaloids and proceed to prove your analysis with the appropriate tests:

1. Brucine and aconite
2. Theobromine and hyoscyamine
3. Cytisine and berberine

## EXPERIMENT 41: THE CHROMATOGRAPHY OF ALKALOIDS

This experiment will use the principles of paper chromatography to identify a series of unknown alkaloids using known $R_f$ factors as a guide. Paper chromatography involves the separation of substances that partition themselves between a liquid stationary phase and a liquid mobile phase. To accomplish this, cut a strip of filter paper the width of a 1000 mL beaker. About 1 inch from the bottom of the paper take a ruler and draw a very light pencil line parallel to the bottom of the paper. Now fill the bottom of the beaker with the solvent mixture called for in the particular experiment. Prior to placing the filter paper down into the solvent, dissolve a few milligram of six different alkaloids in a couple milliliter of ether, and using a small pipet or glass rod place a drop of each dissolved alkaloid along the pencil line, spacing them proportionately. Once you have completed this operation, dry the spots over a warm heater and place the filter paper down into the solvent, making sure the level of the solvent is below the line of test spots. Place a lid on the top of the beaker to create a vapor-tight chamber and wait for the solvent to ascend up the paper, carrying the test spots with it to various heights on the paper. When the solvent is about an inch from the top of the paper, remove from the chamber and allow the paper to dry. For best results with alkaloids, the filter paper should be soaked in a 5% solution of sodium hydrogen citrate and dried at 60° for 25 minutes. The solvent should consist of 50 mL butanol, 50 mL water and 1 g of citric acid. When the paper is dipped into Dragendorff's reagent (5 mL each of solutions A and B) plus 20 mL of acetic acid and sufficient water to make 100 mL, the alkaloids appear on the chromatogram as red spots on an orange background.

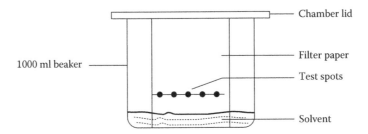

Once the chromatogram has dried you are in a position to measure the total distance moved by your solvent front (the denominator in the fraction). Compare this figure with the distance moved by each of the dissolved alkaloid samples (solute). This gives you the retention factor, usually identified by an "R" on each sample:

$$R_f = \frac{\text{Distance traveled by the solute}}{\text{Distance traveled by the solvent}}$$

The six alkaloids for this experiment are nicotine, brucine, strychnine, atropine, physostigmine, and papaverine. Complete the column for the $R_f$ values and compare with the values given at the end of this section. This will provide you with a partial evaluation of your accuracy in laboratory procedures. Compare results in columns 3 and 4.

| Spot Number | Alkaloid Name | $R_f$ Value | Experimental Value |
|---|---|---|---|
| 1. | Nicotine | _____ | _____ |
| 2. | Brucine | _____ | _____ |
| 3. | Strychnine | _____ | _____ |
| 4. | Atropine | _____ | _____ |
| 5. | Physostigmine | _____ | _____ |
| 6. | Papaverine | _____ | _____ |

### TESTS FOR UNKNOWN ALKALOIDS

In this experiment, you will follow exactly the same procedure as for the previous one, except you will be given six unknown alkaloids whose identity is maintained by your lab instructor. Develop a chromatogram as before and calculate values placing them in the following table.

| Alkaloid Number | $R_f$ Value | Possible Name of Alkaloid |
|---|---|---|
| 1. | | |
| 2. | | |
| 3. | | |
| 4. | | |
| 5. | | |
| 6. | | |

Research and replication of results over the years has led us to be able to establish some fairly well-tuned $R_f$ values for the alkaloids. Using the information from the table below, you may be able to check out the degree of accuracy of your calculations in the first experiment of this section and make a guess as to the identity of the six unknowns in the second.

**$R_f$ Values for Some Select Alkaloids**

| | | | |
|---|---|---|---|
| Nicotine | 0.07 | Cocaine | 0.39 |
| Pilocarpine | 0.11 | Atropine | 0.42 |
| Morphine | 0.12 | Physostigmine | 0.47 |
| Brucine | 0.14 | Papaverine | 0.48 |
| Codeine | 0.16 | Yohimbine | 0.49 |
| Strychnine | 0.25 | Veratrum | 0.82 |

## EXPERIMENT 42: ISOLATION OF ALKALOIDS FROM PLANT SOURCES

The following experiments illustrate the different procedures required to isolate tertiary and quaternary bases. Understanding the cationic salt-forming properties of nitrogenous pharmaceuticals is essential in predicting drug actions, incompatibilities, and product formulations.

### ISOLATION OF HYDRASTINE FROM GOLDENSEAL

Hydrastine is a dark yellow to moderate greenish yellow powder with a distinctive odor and bitter taste and a melting point of 132°. It is freely soluble in acetone and benzine but insoluble in water. It belongs to the isoquinoline class. Its structure is

Ten grams of goldenseal is moistened with sufficient alkaline mixture (ammonia water, ethyl alcohol, and ethyl ether, 1:1:8) to become damp, and the moistened drug is allowed to stand in a covered container for 45 minutes. The material is then transferred to an extraction thimble, and the drug is extracted with ether in a Soxhlet apparatus for at least 6 hours. Following this, the ethereal extract is transferred to a separatory funnel and shaken with three successive portions of 5% aqueous HCl. Evaporate several milliliter of the extracted ether layer, dissolve the residue in 1 mL of 5% HCl, and test it with Mayer's reagent (mercuric-potassium iodide t.s. U.S.P.). If more than a faint precipitate is formed, alkaloids remain in the ether, and the acid extraction should be repeated a fourth time.

The next step is to combine the acid extracts. Make the solution alkaline to litmus with stronger ammonium water and shake it with several portions of ether. Then check 1 mL of the aqueous layer with Mayer reagent for completeness of extraction.

Place the ether extracts in a beaker or flask and remove any water present by filtering through 20 anhydrous sodium sulfate. Place the ether extract in an evaporating disk and evaporate on a steam under the hood. Recrystallize the alkaloid in a small beaker from hot methanol; cool with ice and scrape if necessary.

Dry the crystals on filter paper and weigh: _____.

Determine the melting point of the crystals: _____.

Some color reactions of hydrastine may be illustrated by using the following procedures:

- Place a few crystals of hydrastine in a spot plate and add a few drops of $H_2O_2$ and $H_2SO_4$.
  Result: _____.
- Place a few crystals of the drug in a spot place and add a couple of drops of concentrated $HNO_3$.
  Result: _____.
- Heat a few crystals in a test tube containing 5 mL of concentrated $H_2SO_4$.
  Result: _____.

## ISOLATION OF BERBERINE FROM GOLDENSEAL

A second alkaloid obtained from goldenseal that is of medical importance is berberine, which has the following structure:

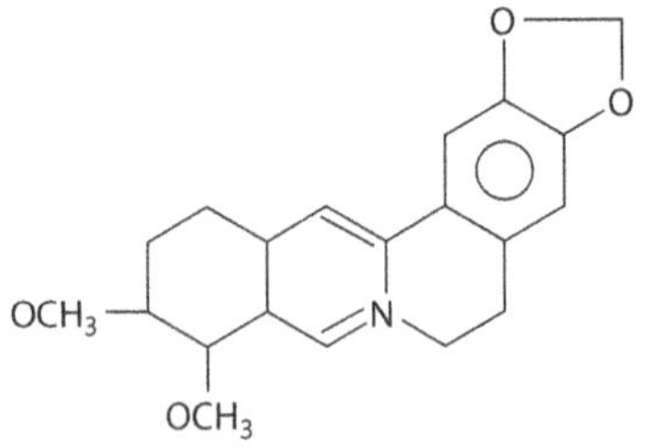

To obtain this drug, carry out the same beginning steps as with hydrastine through removal from the extraction thimble. Spread the powder on a filter paper until the solvents evaporate. Now moisten the powder with ethanol, repack in the thimble, and extract for about 5 hours. Place your portion of the alcoholic extract in a beaker and evaporate over a hot plate until near dryness. Add 50 mL of $H_2O$ to the residue and boil to dissolve the berberine. Filter the solution while it is still hot through cotton place in a funnel. Add 5 mL of 5% HCl until the solution is acidic and allow the solution to cool. Berberine HCl will settle as dark crystals. Filter off the crystals, dry, and weigh. Weight of berberine _____.

Determine the melting point of the berberine _____.

Some color reactions of berberine may be shown by the following:

- Place a few crystals of the drug in a test tube containing 5 mL of concentrated $H_2SO_4$ and heat.
  Result: _____.
- Place a few crystals in a spot plate with concentrated $HNO_3$.
  Result: _____.

  (Note difference in reaction s between hydrastine and berberine.)

## ISOLATION OF CAFFEINE FROM TEA

Caffeine belongs to the purine class of alkaloids and is present in coffee, lea, guarana paste, and cola nuts. It is soluble in water, alcohol, acetone, chloroform, benzine, and ether. Solubility in water is increased by the presence of alkali benzoates, cinnamates, citrates, or salicylates. The structure of caffeine is

Place 50 g of powdered tea and 250 mL of water in a 600 mL beaker and boil gently for 15 minutes. Strain the resulting hot extract through muslin. Carefully add a solution of basic Pb $(C_2H_3O_2)_2$ to the filtrate until no more precipitate forms. Heat the mixture to boiling and filter through a Büchner funnel and vacuum system. The filtrate is heated to boiling and dilute sulfuric acid is added until precipitation occurs. Approximately 1 g of decolorizing charcoal is added to the mixture, which is boiled for a few minutes and then filtered. The filtrate should he collected in a large evaporating dish. One gram of decolorizing charcoal is again added to the filtrate, which is then evaporated to a volume of 100 mL.

Cool the mixture and transfer to a separatory funnel and shake out with three successive 10 mL portions of $CHCL_3$. Transfer the combined extracts to a small evaporating dish and warm to dryness. Scrape out the residue, transfer to a small beaker, and dissolve in a small amount of hot $CH_3OH$. Let stand overnight and filter off the caffeine crystals.

Record weight of the caffeine _____.

Record melting point of the caffeine _____.

Check with *Merck Index* for your accuracy.
Some color reactions of caffeine are

- Dissolve 5 mg of caffeine in 1 mL of HCl, add 50 mg KClO$_3$, and evaporate on a steam bath to dryness. Invert the dish over a vessel containing ammonia t.s.
  Result: _____.
- To 5 m of a saturated solution of caffeine add 5 drops of iodine t.s. Then add 3 drops of diluted HCl.
  Result: _____.

## ISOLATION OF STRYCHNINE FROM NUX VOMICA

Strychnine and its relative brucine are both members of the indole class of alkaloids. Their structures are:

Strychnine　　　　　　　　　　　　　　Brucine

Isolation procedure begins with 8 g of nux vomica powder being placed in a flask containing 25 chloroform, 50 mL of ether, and 5 mL of 10% ammonia water. The mixture should be shaken frequently during a half-hour period, and then 50 mL of the solution is filtered through cotton and transferred to a separatory funnel. Add dilute H$_2$SO$_4$ and shake well. The lower aqueous layer, which should have an acid reaction when tested with litmus paper, is drawn off into a second separatory funnel and the eth chloroform mixture extracted twice more with dilute acid.

The mixed acid extractions containing the alkaloids from the original drug should now be render alkaline with dilute ammonia water, and the liberated alkaloidal bases completely extracted by shaking with several portions of chloroform. After the chloroform extractions are washed with a little water, solvents may then be driven off by heating in an evaporating dish. A little ethyl alcohol is added and once evaporated, and the residue, consisting of strychnine and brucine, is dried at 100° and weighed.

Record the weight of the alkaloidal mixture _____.

The process for separating strychnine from brucine depends on the greater readiness by which brucine is nitrated with HNO$_3$. Dissolve the alkaloidal mixture in 10 mL of dilute HSO$_4$. Using 50 mL of H filter into a graduated cylinder. Pour the cylinder contents into a flask and add exactly 5 mL of HNO$_3$ concentrate. The addition of the HNO$_3$ should cause the solution to assume a brilliant crimson color. After standing for exactly 15 minutes, the liquid is transferred to a separatory funnel and at once rendered alkaline with NaOH solution. The strychnine is extracted with three portions of chloroform in the usual way. After boiling off the chloroform in an evaporating dish, a little ethyl alcohol is added and evaporated to dryness. After drying at 100°, the residue of strychnine is weighed.

Result: _____.

Color tests are available to differentiate between strychnine and brucine. Conduct the following tests and record the results: To 0.1 g of strychnine add 1 mL of a 50:50 mixture of concentrated $HNO_3$ and $H_2O$, and record the result of the spot test: _____.

Perform the same test with 0.1 g of brucine.

Result: _____.

Now add a few drops of the water to the brucine and add a few drops of stannous chloride T.S.

Result: _____.

Check the NF or USP/NF to validate the accuracy of your results.

## EXPERIMENT 43: ASSAY OF BELLADONNA FOR HYOSCYAMINE

Alkaloidal assays are commonly performed for purposes of standardization, proof of purity, commercial evaluation, or pharmacolegal purposes. A slight deficiency of alkaloid in a preparation may cause a marked decrease in the physiological effect, or a slight excess may cause toxic effects when the preparation is administered.

The amounts of alkaloids that occur in crude drugs are subject to considerable variation in different samples of the same drug. The variations may be caused by

1. The age of the plant when it is collected
2. The season of the year when the drug is harvested
3. The soil and climate in which the drug is grown
4. The conditions under which the drug is collected, stored, and dried

In view of the fact that alkaloids may comprise only a fraction of 1% of the substance assayed and that this small amount must be separated from numerous other constituents present in the crude drug, such as resins, volatile oils, coloring matter, glycosides, fatty acids, gums, and proteins, it is evident that the exact technique involved in any given method must be carefully adhered to in order to estimate the variations in alkaloidal content. It is this special technique that characterizes the chemical assay of drugs, rather than the gravimetric or volumetric nature of the procedure employed.

### Aklaloidal Determination in Belladonna

Place 10 g of belladonna powder in a dry flask and macerate with 10 mL of ethyl alcohol and 20 mL of ether for 10 minutes, then add 5 mL of ammonia water to render the solution alkaline and liberate the alkaloidal bases from their salts. After allowing to stand with frequent shaking for an hour, transfer the contents of the flask to a small percolator as shown in the diagram. Percolate the drug first with another ether–alcohol mixture and then with ether alone, until the alkaloids are extracted. In order to ascertain when complete extraction has taken place, a few drops of the solvent is collected on a watch glass and the solvent evaporated. Given any residue with dilute HCl and a couple of drops of Mayer's reagent, if extraction is incomplete, a cream-colored precipitate or turbidity indicative of the presence of an alkaloid will appear. Once the liquid remains clear, extraction is complete. Owing to the tendency of the alkaloids to hydrolyze during extraction, the percolation should not last more than three hours.

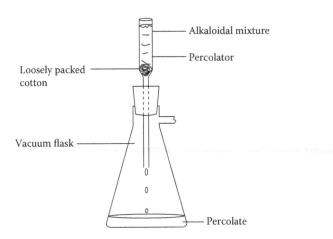

The percolate should now be transferred to a separatory funnel and shaken with an excess of dilute HCl. After separation, the lower level is drawn off into another separatory funnel and the ethereal layer twice more extracted with small portions of dilute HCl mixed with a little alcohol. In this way, the alkaloids are transferred from there to the aqueous acid liquid. The alcohol is added to prevent the formation of troublesome emulsions. The acid solution of the alkaloids is now freed from traces of chlorophyll and, extractive matter with 10 mL of chloroform, allowing to separate and drawing off the chloroformic liquid into another separatory funnel and then shaking with a little dilute acid in order to remove any traces of alkaloids that may have passed into the chloroform.

After separation, the chloroform is rejected, and the acidic solution now rendered alkaline with 10 mL of ammonia water. This liberates the alkaloidal bases, which are sparingly soluble in water, and they can then be extracted by shaking with several successive portions of chloroform. Two milliliter of ethyl alcohol should be added to the residue, then evaporated off a water bath, and the residual alkaloid dried at 100°C.

The alkaloidal residue is now dissolved in 10 mL of .02 N HCl and the excess titrated with .02 N NaOH employing methyl red as an indicator. Each milliliter of .02 N HCl used is equivalent to 0.005784 g of alkaloid calculated as hyoscyamine.

Result: _____.

The structure for hyoscyamine is

## VITALI'S IDENTIFICATION TEST

Place two drops of concentrated $HNO_3$ on a few milligram of a solanaceous alkaloid (atropine, hyoscyamine, scopalamine, or strammonium), and on another evaporating dish a few milligram of homatropine. Evaporate to dryness. Now moisten the various residues with a few drops of alcoholic KOH solution.

Result: _____.

Explain the reason the homatropine reacted differently than the other samples in this group.

## EXPERIMENT 44: ASSAY OF AMINOPHYLLINE TABLETS

Aminophylline finds use in pharmacology as a bronchodilator. Its veterinary uses include being a smooth-muscle relaxant and as a diuretic in dogs who are suffering from congestive heart failure. The chemical name for aminophylline is 3, 7 dihydro-1, 3 dimethyl 1-H purine- 2.6 dione plus 1.2 ethenediamine. Its formula is $C_{16}H_{24}N_{10}O_4$ $2H_2O$.

*Procedure for the assay*: Weigh out 20 aminophylline tablets and grind them with a mortar and pestle to a powder. The tablets are 3 grains each, making a total of 60 grains (about 4.0 gram). Transfer the powder to a 200 mL flask and add 50 mL pf water and 15 mL of ammonia t.s. and place on a magnetic stirrer for 30 minutes. Then dilute to 200 mL with water, mix well, and filter into a dry flask, rejecting the first 20 mL of the filtrate. Transfer an accurately measured aliquot, equivalent to about 300 mg of aminophylline to a 250 mL Erlenmeyer flask and add water, if necessary, to make 40 mL.

The next step is to add 8 mL of ammonia t.s. and 20 mL of $AgNO_3$ (0.1 M) mix and warm on a steam bath for 15 minutes. Cool the mixture to between 5° and 10°C for 20 minutes and filter using a vacuum pump, then wash the precipitate three times with 10 mL portions of water.

Now you can acidify the combined filtrate and washings with $HNO_3$. Cool the mixture and add 2 mL of ferric ammonium sulfate t.s. and titrate the excess $AgNO_3$ with 0.1 M ammonium thiocyanate ($NH_4SCNO$).

Each milliliter of 0.1 M $AgNO_3$ is equivalent to 22.82 mg of aminophylline. Record results:

_____mL of $AgNO_3$ × 22.82 _____mg $C_{16}H_{24}N_{10}O_4$ 2 $H_2O$

Calculate the percentage of the labeled amount of hydrated theophylline in the tablets assayed.

## SELECTED REFERENCES—ALKALOIDS

Bentley, K. *The Alkaloids*. Wiley, 1957.
Cassiano, N. *Alkaloids: Products, Application, and Pharmacological Effects*. Nova Science, 2011.
Daniel, M. *Medicinal Plants*. CHIPS, 2006.
Feigl, F. *Spot Tests in Organic Analysis*. Elsevier, 1959.
Haas, P. *Chemistry of Plant Products*. Longmans, 1928.
Heinrich, M. *Fundamentals of Pharmacognosy*. Elsevier, 2011.
Hesse, M. *Alkaloid Chemistry*. Wiley, 1981.
Ikan, R. *Natural Products Lab Guide*. Academic Press, 1969.
Knoller, H. J. *The Alkaloids*. Springer, 2011.
Orazio, T. S. *Modern Alkaloids*. VCH Publishing, 2010.
Patrick, G. *An Introduction to Medicinal Chemistry*. Oxford University Press, 2009.
Robbers, J. *Pharmacognosy Lab Guide*. Purdue University Press, 1981.
Rosenthaler, L. *Chemical Investigation of Plants*. Bell, 1930.

# 11 Tannins

## TANNINS

Tannins comprise a large category of substances widely distributed in the plant kingdom. They are noncrystallizable chemicals when mixed with water and form colloidal solutions possessing an acid reaction and an astringent effect. They usually occur as mixtures of polyphenols.

Tannins precipitate proteins from solution and are able to combine with them, rendering them resistant to proteolytic enzymes. When applied to living tissues this action is known as *astringent*. Tannins are classified as

1. catechols, such as gambir and catechu or
2. pyrogallotannins—found in nutgall and oak bark.

Tannins find use in medicine in such preparations as witch hazel extract used as an astringent to relieve the pain of bruises and hemorrhoids and tannic acid used in ointments as an astringent.

Experiments in this section illustrate color reactions of selective tannins, including those in nutgall, calumbra, and oak bark.

## EXPERIMENT 45: COLOR REACTIONS OF SELECTIVE TANNINS

Tannins are complex substances usually occurring as polyphenols and often difficult to separate because they do not crystallize. Recent chromatographic work has, however, alleviated part of the problem and allowed researchers to identify the simple polyphenols present in mixtures. The tannins may be divided into two classes:

1. Hydrolyzable, consisting of gallic acid or related polyhydric compounds esterified with glucose. Because such esters are readily hydrolyzed to yield the phenolic acids and the sugar, they are referred to as *hydrolyzable*. The structure of gallic (3,4,5 tri-hydroxy benzoic) acid is

2. Nonhydrolyzable, containing only phenolic nuclei, are frequently linked to carbohydrates or proteins. When treated with hydrolytic agents, they tend to polymerize yielding, usually red compounds known as phlobophenes.

Both classes of tannins are widely distributed in nature. Some examples are such substances as hamamelis leaf (witch hazel) and nutgall, which have particular pharmaceutical significance as astringents and for the treatment of burns.

Other usable ways to distinguish tannins is to divide them into categories of phlobotannins, which yield catechol and give a green reaction with ferric salts, and pyrogallotannins, which yield pyrogallol and give a blue color test with ferric salts.

Tannins are easily identifiable by color tests. In this experiment, you will set up nine test tubes of about 40 mL capacity, labeled 1 through 9. The next step is to boil 1 g of nutgall in 50 mL of water for 5 minutes. Cool and filter. Dilute 1 mL of the extract with 10 mL of water and place in test tube 1. Follow the same dilution procedure and place 1 mL of the nutgall extract in tubes 2 and 3. Next, follow the same extraction, filtering, and diluting procedures for calumba root powder, which will be placed in tubes 4, 5, and 6, and with oak bark, which will be placed in tubes 7, 8, and 9.

In tubes 1, 4, and 7, place 10 drops of a 1% solution of $FeCl_3$. In tubes 2, 5, and 8, place 10 drops of 1% solution of $Pb(C_2H_3O_2)$, and in tubes 3, 6, and 9, place 10 drops of a 1% solution of quinine.

Record all observations in the table below, making particular note of color reactions and precipitate occur. Make an estimate of the class of tannins that predominate in each of the extracts, keeping in the facts that most naturally occurring substances are mixtures of both classes.

| Extract | Reagent | Observation | Class |
|---------|---------|-------------|-------|
| 1. Nutgall | $FeCl_3$ | | |
| 2. Nutgall | $Pb(C_2H_3O_2)_2$ | | |
| 3. Nutgall | Quinine | | |
| 4. Calumba | $FeCl_3$ | | |
| 5. Calumba | $Pb(C_2H_3O_2)_2$ | | |
| 6. Calumba | Quinine | | |
| 7. Oak bark | $FeCl_3$ | | |
| 8. Oak bark | $Pb(C_2H_3O_2)_2$ | | |
| 9. Oak bark | Quinine | | |

Some plants with tannin content of particular importance include:

| Botanical Name | Common Name | Tannin |
|----------------|-------------|--------|
| *Castanea dentata* | Chestnut | Tannic acid |
| *Hamamelis virgiana* | Witch hazel | Hamamelitannin |
| *Krameria trianda* | Rhatany | Catechol |
| *Quercus infectoria* | Nutgall | Tannic and gallic acids |
| *Rosa gallica* | Red rose | Pyrogalloltannin |
| *Uncaria gambir* | Gambir | Catechutannic acid |

The structure of catechin is

## SELECTED REFERENCES—TANNINS

Allport, N. *The Chemistry and Pharmacy of Vegetable Drugs*. Newnes, 1945.
Barton, D. *Comprehensive Natural Products Chemistry*. Pergamon, 1998.
Breitman, E. *Terpenes*. Wiley, 2006.
Hagerman, A. *The Tannins Handbook*. Hagerman, 2011.
Haslam, E. *Comprehensive Natural Products Chemistry*. Pergamon, 1999.
Haslam, E. *Chemistry of Vegetable Tannins*. Pergamon, 1966.
Hemingway, R. *Chemistry and Significance of Condensed Tannins*. Plenum Press, 1989.
Hemingway, R. *Plant Polyphenols*. Springer, 1992.
Howes, F. N. *Vegetable Tanning Materials*. Chronica Botanica, 1953.
Nirenstein, M. *The Natural Organic Tannins*. Churchill, 1934.
Quideau, S. *Chemistry and Biology of the Ellagitannins*. World Scientific, 2009.
Robbers, J. *Pharmacognosy Laboratory Guide*. Purdue University Press, 1981.

# 12 Resins

## RESINS

Resins are complex mixtures of resin acids, resin alcohols, resinotannols, esters, and resenes. They are insoluble in water but dissolve in alcohol and other organic solvents.

Resins may occur as

1. Oleoresins—homogeneous mixtures with volatile oils
2. Gum resins—mixtures with gums
3. Balsams—resin mixtures with cinnamic and benzoic acids
4. Glycoresins—found with glycoside combinations

Some resinous products are used in pharmacy. Podophyllum resin is sometimes used as a caustic for papellomas.

Colocynth was listed in the National Formulary as an intestinal purgative although it is not used much today due to its drastic activity. The same is true for jalap root, cannabis, or American hemp, which has found some use as a treatment for glaucoma in the form of tetrahydrocannabinol (THC). The resin of the cayenne pepper, or capsicum, finds use as a carminative, rubefacient, and stimulant.

The experiments on resins illustrate the properties of some resins. A comparative study of African and Jamaican ginger, which are two related but distinct types of resins occurring in different areas of the world, has been conducted.

## EXPERIMENT 46: PROPERTIES OF SELECTED RESINS

Resins are amorphous products with a complex chemical nature. They are usually found in schizogenous or schizolysigenous ducts or cavities and are the end products of metabolism. They are complex mixtures of resin acids, resin alcohols, resinotannins, esters, and resenes. Numerous resins are used in medicine and pharmacy, including rosin, guaiac, mastic, ginger, tolu balsam, Peruvian balsam, and benzoin. They are divided into four basic categories: oleoresins, oleo-gum-resins, balsams, and glycoresins.

Examples in each category include

Oleoresins—turpentine, copaiba, ginger, and capsicum
Oleo-gum-resins—asafetida and myrrh
Balsams—benzoin, tolu, peru, and styrax
Glycoresins—jalap and podophyllum

The first activity in this section will be to prepare jalap resin from the crude jalap powder root. Place 5 g of the powder in a 250 mL Erlenmeyer flask and add 100 mL of a 50:50 mixture of methanol (CH$_3$OH) and water. Swirl to mix thoroughly. Next, set up a percolation apparatus as shown and percolate the jalap powder solution through the cotton into a receiving beaker. Place the beaker in a water bath and boil gently until you have concentrated the percolate to one fourth its original volume.

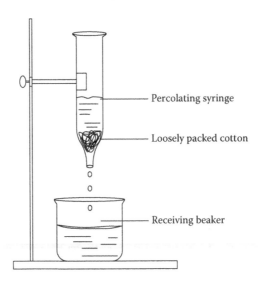

Next pour the concentrated percolate into hot water. This action results in a precipitate of jalap resin. Filter, dry, and weigh. Calculate percentage of resin in the crude powder.
Calculation:
Some characteristic color reactions can be achieved with some of the resins. Mix 0.5 g of podophyllum powder in 10 mL of NaOH T.S. Record the color.
Result: _____.
Another color reaction can be illustrated using tincture of myrrh. Place 5 mL of tincture of myrrh in a small test tube and add 2 mL of concentrated HNO$_3$.
Color result: _____.
Place 5 mL of tincture of myrrh in a small test tube and add 2 mL of bromine water.
Color result: _____.
Check the USP/NF for the accuracy of your results.
Water-soluble extractives are also an important part of resin analysis. Place 4 g of powdered ginger in a 200 mL volumetric flask. Fill to the mark with water and agitate at 30-minute intervals for 8 hours of slowly agitate on a magnetic stirrer for the same period of time. Allow the mixture to stand for 16 hours and filter. Evaporate 50 mL of the filtrate, representing 1 g of the drug, on a water bath and dry the residue at l05°C for 2 hours. Weigh the extract and record.
Results: _____.
In addition to water extractives, alcohol extractives are an important measurement in resin studies. Place 2 g of asafetida in a tared extraction thimble and extract with C$_2$H$_5$OH in a Soxhlet extraction apparatus for 3 hours. Dry the residue at 105°C for 2 hours and weigh.
Record results: _____.

Next determine the moisture in the drug by the toluene distillation method using the apparatus shown:

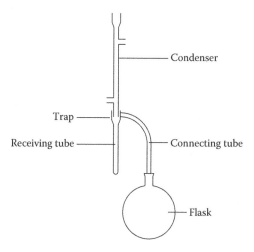

Place 5 g of asafetida in the flask along with 200 mL of toluene and connect the apparatus. Fill the receiving tube with toluene and begin to boil. Distill at the rate of about two drops per second until most of the water has passed over. Then increase the distillation to about four drops per second. When the water has distilled over, rinse the inside of the condenser tube with toluene while brushing down the tube with a tube brush attached to a copper wire and saturated with toluene. Continue the distillation for 5 minutes, then remove the heat and allow the receiving tube to cool to room temperature. If any droplets of water adhere to the walls of the receiving tube, force them down with a rubber band wrapped around a copper wire and wet with toluene. When the water and the toluene have separated completely, read the volume of the water and calculate the percentage that was present in the asafetida. Subtract this weight of moisture from the original weight of the asafetida. The difference between this result and the weight of the residue determined above represents the alcohol-soluble extractive.

Calculations:

The final activity in this experiment section on resins will be to determine the ash content percentage of capsicum (red or cayenne pepper). Accurately weigh 4 g of capsicum powder in a tared crucible and add 15 mL of $CH_3OH$. Triturate the powder in the alcohol with a glass rod. Next incinerate, using a Bunsen burner at a temperature not to exceed dull redness. Cool and determine the weight of the ash. Calculate the percentage of total ash and record results.

Calculations:

$$\frac{(\quad)g - \text{weight of the ash}}{4 \text{ g} - \text{orginal of leaves}} \times 100 = \underline{\quad} \%$$

# EXPERIMENT 47: COMPARATIVE STUDY OF AFRICAN AND JAMAICAN GINGER

There are numerous forms of ginger available commercially that are generally named according to their region of origin. We have African, Jamaican, Cochin, Japanese, and Martinique among others, but only the first two have found their way into pharmaceutical practice.

Ginger has a very long history of use and was known in China as early as the 4th century B.C. It was used as a spice by the Greeks and Romans and from the 11th through the 13th centuries was a common import from the East. It was introduced into Jamaica by Spanish colonists in the 16th century, and this has resulted in its being an important commercial product of that island ever since.

Gingers consist of a volatile oil (1% to 3%) that gives rise to their aroma and a viscid oily resinous liquid known as gingerol (0.5% to 1.5%) that creates its pungency. There are also other resins, starches, mucilages, and homologous phenols present.

In this experiment, we will conduct tests on water-soluble extractives and ether-soluble extractives, and compare results between African and Jamaican ginger.

## WATER-SOLUBLE EXTRACTION

Place 4 g of ground African ginger in a 200 mL volumetric flask and fill with water. Agitate, using a magnetic stirrer for 8 hours. Filter and evaporate 50 mL of the filtrate on a water bath and dry the residue at 105°C for 2 hours. Weigh the extract and record. Follow the same procedure with Jamaican ginger and record its water-soluble extractive.

African ginger water-soluble extractive _____
Jamaican ginger water-soluble extractive _____

## ETHER-SOLUBLE EXTRACTION

Place 20 g of African ginger in an extractive thimble of a Soxhlet or other extractive apparatus and extract with 100 mL of ether for 6 hours. Evaporate the ether on a steam bath. Then dry the residue in a dessication over $H_2SO_4$ for 18 hours. Weigh the extract. Repeat this procedure using the Jamaican ginger and recon results.

African ginger ether-soluble extractive _____
Jamaican ginger ether-soluble extractive _____

As a final project in this experiment, check the USP/NF and other pharmaceutical literature and list some preparations that make use of ginger.

## SELECTED REFERENCES—RESINS

Atkinson, E. *Gums and Gum Resins*. Bibliolife, 2010.
Bernfield, F. *Biogenesis of Natural Compounds*. Pergamon, 1967.

Coppen, J. J. W. *Gums, Resins, and Latexes of Plant Origin.* Food & Agricultural Organization, 1995.

DeMayo, P. *Chemistry of Natural Products.* Interscience, 1959.

Gokhale, S. B. *Pharmacognosy.* Pragmatic Books, 2008.

Guenther, E. *The Essential Oils.* Van Nostrand, 1948–1952.

Howes, G. N. *Vegetable Gums and Resins.* Chronica Botanica, 1949.

Langenheim, J. *Plant Resin Chemistry.* Timber, 2003.

Ramstad, E. *Modern Pharmacognosy,* McGraw-Hill, 1957.

Salter, C. *Analysis of Resins, Balsams, and Gum Resins.* Bibliolife, 2010.

# 13 Glycosides

## GLYCOSIDES

Glycosides are compounds that yield, upon hydrolysis, one or more sugars. Chemically, they are acetals in which the hydroxyl of the sugar is condensed with a hydroxyl group of the nonsugar, and the secondary hydroxyl is condensed with the sugar molecule itself to form an oxide ring.

Glycosides are classified as

1. Cardioactive such as digitalis
2. Anthroquinone such as *Cascara sagrada*
3. Saponin such as glyyrrhiza
4. Cyanophone such as wild cherry
5. Thiocyanate such as mustard
6. Flavone such as hesperidin
7. Alcohol such as salicin
8. Aldehyde such as vanilla
9. Lactone such as coumarin
10. Phenol such as uva ursi
11. Others such as gentian

Experiments in this section include the synthesis of some glycosides, color reactions of different types of glycosides, some analytical tests to differentiate some types of glycosides, and a specific study of the cardiac glycoside digitalis.

## EXPERIMENT 48: GLYCOSIDE SYNTHESIS OF EMULSIN

Glycosides are compounds that yield one or more sugars among the products of hydrolysis. The most frequently occurring sugar is $\beta$-D-glucose, although rhamnose, digitoxose, cymarose, and other sugars are also components of glycosides. When the sugar formed is glucose, the substance may be called a glucoside. However, because other sugars may be developed during the hydrolysis, the term *glycoside* is used.

Chemically, the glycosides are acetals in which the hydroxyl of the sugar is condensed with a hydroxyl group of the nonsugar component, and the secondary hydroxyl is condensed within the sugar molecule itself to form an oxide ring. An example is the glycoside prunasin:

While glycosides are extremely difficult to categorize due to their complex and variant structure, the most workable current scheme seems to be the following:

1. Cardioactive—so-called because of their action on the cardiovascular system. They include digitalis, deslanoside, convallaria, apocynum, and squill.
2. Anthraquinone—usually employed as cathartics and include cascara, frangula, aloe, senna, rhubarb, and chrysarobin.
3. Saponin—from colloidal solutions that usually foam in water and include glycyrrhiza (licorice) and dioscorea (yam).
4. Cyanophore—yield hydrocyanic acid as one of the by-products and include wild cherry and amygdalin.
5. Isothiocyanate—when hydrolyzed by the enzyme myrosin, yield mustard oils and include *Sinapis nigra* (black mustard) and *Sinapis alba* (white mustard).
6. Flavonol—include many natural yellow pigments such as rutin, quercetin, and hesperidin.
7. Alcohol glycosides—hydrolyzed into alcohols and include salicin, which is obtained from willow bark.
8. Aldehyde glycosides—have an aldehydic aglycone as chief constituent and include vanillin, used as a flavoring agent.
9. Lactone glycosides—include coumarin, santonin, and cantharides (Spanish fly).
10. Phenol glycosides—have a phenolic ring or hydroxy group on the aromatic system and include arbutin in uva ursi (bearberry) and phloridzine in the bark of rosaceous plants.
11. Tannins—a special category of glycosides that will be covered in detail in another section.

To perform the experiment on the synthesis of emulsin, prepare a solution of $(NH_4)_2SO_4$ by dissolution 56.5 g in 150 mL of water. Add 100 mL of this solution to 10 g of defatted bitter almond seeds. Stir 5 minutes, filter, and discard the filtrate. Next, extract the meal again, with 50 mL of the $(NH_4)_2SO_4$ solution diluted with 50 mL of water. Add 10 g of $(NH_4)_2SO_4$ to this filtrate, collect the precipitated emulsion, and suspend the material in 50 mL of acetone. Stir for minutes and filter. Dry and maintain the product in a dessicator. Weigh the yield.

Result: _____.

The second portion of this experiment is to dissolve 100 mg of salicin in 10 mL of water contained in a test tube. Then add 50 mg of emulsin to the solution and shake thoroughly. Remove 1 mL of the solution at zero time and subsequently after 5, 10, 20, and 30 minutes. Place the withdrawn samples in a plate and immediately determine the glucose with a small piece of Testape. Then add a drop of $FeCl_3$ to the solution and observe the colors produced. Record results and explain the observed changes.

| Time (Minutes) | | Color |
|---|---|---|
| 0 | _____ | _____ |
| 5 | _____ | _____ |
| 10 | _____ | _____ |
| 20 | _____ | _____ |
| 30 | _____ | _____ |

## GLYCOSIDE (ACETAL) FORMATION

| Monosaccharid3 | Alcohol | Glycoside |

| Monosaccharide | Monosaccharide | Disaccharide, a glycoside |

## EXPERIMENT 49: SOME COLOR REACTIONS OF GLYCOSIDES

In this experiment, you will perform some tests that produce identifying color reactions by selected glycosides: cascara sagrada, an anthraquinone; vanillin, an aldehyde; chrysarobin, an anthraquinone; and santonin, a lactone.

For the first test, add 100 mg of powdered *Cascara sagrada* to 10 mL of hot water. Shake the mixture gently until it cools. Filter, dilute the filtrate with water to 10 mL, and add 10 mL of ammonia t.s. Record the color produced.

Result: _____.

In the second test, to 10 mL of a cold, saturated solution of vanillin, add 5 drops of $FeCl_3$ t.s. Record color.

Result: _____.

Now heat this mixture at 80°C for a few minutes and record color change.

Result: _____.

After cooling, there is a third color change.

Result: _____

The third test is with chrysarobin. Mix 0.1 g of chrysarobin with 5 drops of fuming $HNO_3$. Note color.

Result: _____.

Now add 5 drops of ammonia t.s.

Result: _____.

The final color test of this section will use santonin, a lactone, with the structural formula:

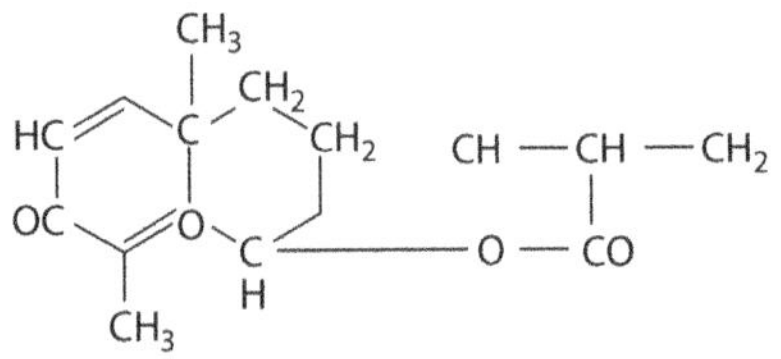

Heat 0.2 g of santonin with 2 mL of alcoholic KOH t.s. and record the color reaction.

Result: _____.

Next shake 10 mg of santonin with a cool mixture of 1 mL each of $H_2SO_4$ and distilled water. Heat solution to 100°C and add a drop of $FeCl_3$ t.s. diluted to 10 mL with distilled water. Record the reaction.

Result: _____.

Finally, in order to check the accuracy of your work, refer to a USP or NF, which indicate the results you should get with each of these procedures.

## EXPERIMENT 50: DIFFERENTIATION OF CHRYSAROBIN AND GLYCOSIDES

From 1882 to the present, the U.S. Pharmacopeia has included chrysarobin in its listings. Chrysarobin is a reddish-brown powder obtained from the wood of the *Andira araroba* tree that grows in the Indian state of Goa, a former Portuguese colony on the Malabar Coast, and also in the provinces of Bahia and Sergipe in Brazil.

Its basic structure is classified as an anthraquinone, and, as such, it shares physiological action properties with cascara, rhubarb, senna, and aloe. The purpose of this exercise will be to differentiate chrysarobin from other glycosidal irritants using various reagents that yield specific reactions.

The molecular structure of chrysarobin is:

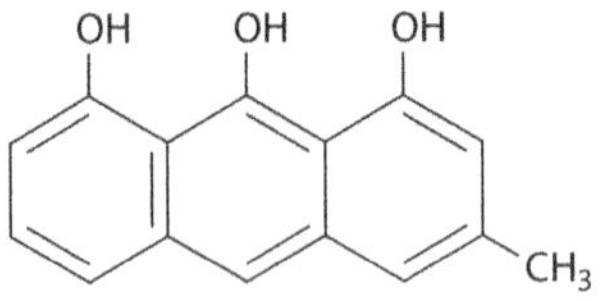

$(C_{15}H_{12}O_3)$

*Procedure*: Place 0.05. g of chrysarobin, rhubarb, aloe, and senna in four different test tubes numbered 1 through 4. Add 1 mL of fuming $HNO_3$ to each tube and record results.

Results with fuming $HNO_3$:

1 _____

2 _____

3 _____
4 _____

Set up four test tubes again, placing each of the four just mentioned substances (0.05 g) in separate tubes. This time add 1 mL of ammonia t.s. (10% NH$_4$OH) and record your results

Results with ammonia:

1 _____
2 _____
3 _____
4 _____

Check the results obtained using the USP/NF as a resource. Record any differences with the information given in this reference.

Check REMINGTON'S and the USP/NF for information on the pharmaceutical uses of these substances and list below.

## EXPERIMENT 51: ALOES IDENTIFICATION TESTS

There are about 150 species of aloe known most of which are indigenous to Africa, although some have been introduced in the West Indies and Europe. They are classified as glycosides. Glycosides are compounds that yield one or more sugars among the products of hydrolysis. The sugar component is known as a glycone, and the nonsugar component is an aglycone.

Chemically, the glycosides are acetals in which the hydroxyl group of the sugar is condensed with a hydroxyl group of the nonsugar component, and the secondary hydroxyl is condensed within the sugar molecule itself to form an oxide ring. Both alpha and beta glycosides are possible. Thus we have

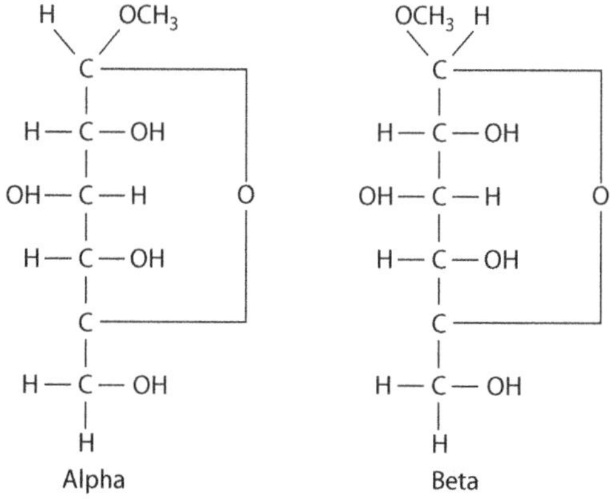

In this experiment, we are going to differentiate 3 major species of aloe which each give distinct and different results. The 3 types are *Aloe perryi*, known as

socotrine aloe; *Aloe barbadensis*, known as Curacao aloe; and a hybrid of *Aloe africana* and *Aloe spicata*, known as Cape aloe.

*Procedure*: Place 1 g of each sample of the aloe in separate 100 mL beakers. Add 25 of cold water and place on a magnetic stirrer for two hours. Next, filter each of the three samples. Then place the filtrate and 100 mL of water in volumetric flasks and swirl vigorously for two minutes. Record results:

Color of socotrine aloe filtrate _____
Color of Curacao aloe filtrate _____
Color of Cape aloe filtrate _____

To 5 mL of each of three filtrates from above, add 2 mL $HNO_3$ and record results:

Color of socotrine aloe/acid mix _____
Color of Curacao aloe/acid mix _____
Color of Cape aloe/acid mix _____

Finally, check the accuracy of your results in a natural products or pharmacognosy text.

## EXPERIMENT 52: THE CARDIAC GLYCOSIDES OF DIGITALIS

Perhaps the most famous and well-known of all the glycosides are the various cardiac glycosides, so-called because of their action on the heart muscles. Some of their structural formulas shown below indicate a basic cyclophenanthrene nucleus to which is attached a lactone ring.

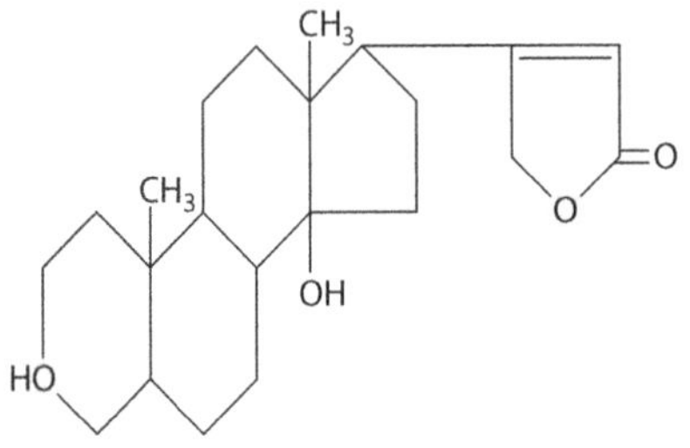

Digitoxigenin

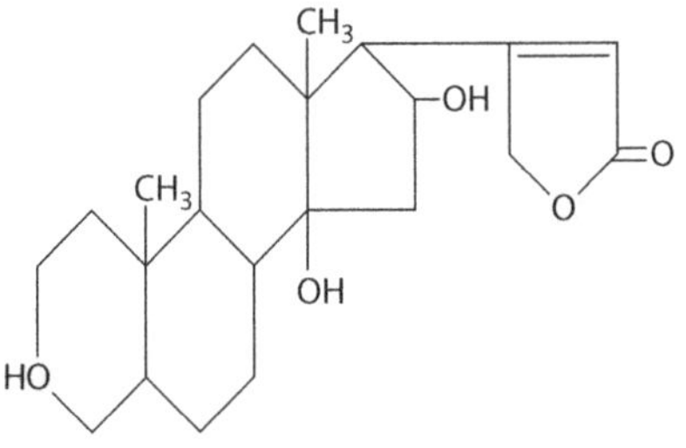

Gitoxigenin

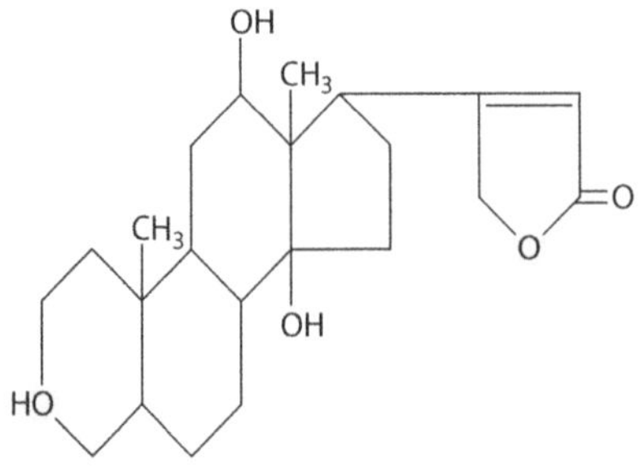

Digoxigenin

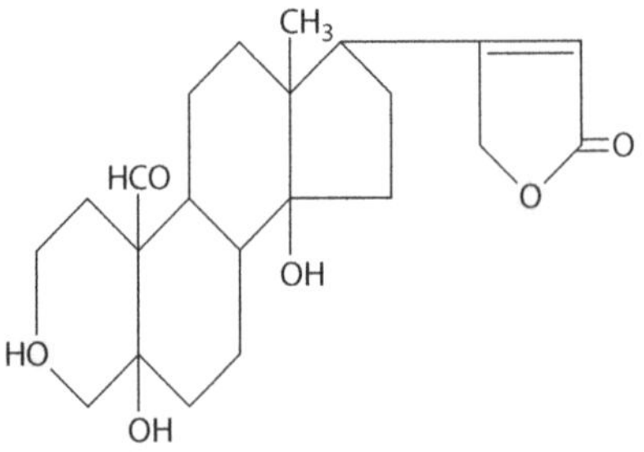

Strophanthidin

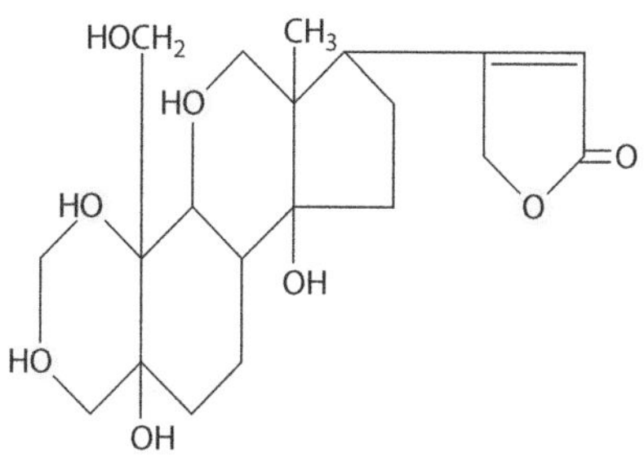

Ouabagenin 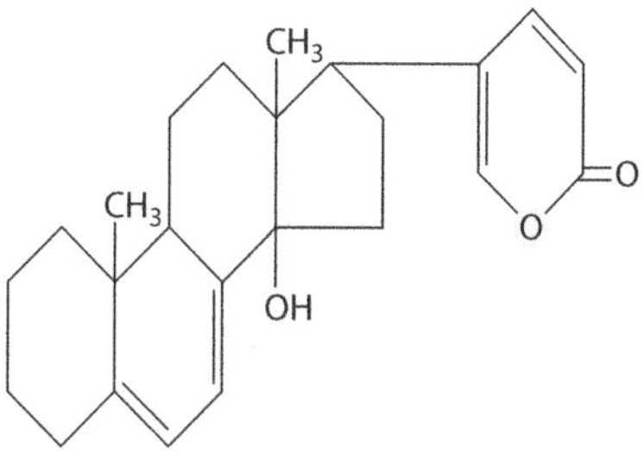                                      Scillaridin A

Digitalis is practically insoluble in water, but there are two tests that can be used to identify it.

Reaction with ethanol—Place 1 g of digitalis leaves in 50 mL of concentrated HCl.

Record color result _____.

Keller's reaction: Dissolve 1 g of digitalis leaves in 10 mL of glacial acetic acid in a large test tube. Next, add a drop of ferric chloride solution and then, gently some $H_2SO_4$ to form a layer below the acetic acid.

Describe the color reaction which occur:

Determining the ash content of digitalis leaves. Accurately weigh 4 g of the leaves and place in a tared crucible. Then incinerate at a low temperature not to exceed very dull redness. Cool the crucible and ash in a dessicator and weigh. Record results:

$$\frac{\text{Weight of ash}}{\text{Weight of leaves}} \times 100 \underline{\hspace{1cm}} \%$$

Determining acid-insoluble ash content of digitalis. Boil the ash obtained from the previous procedure with 25 mL of dilute HCl for 5 minutes. Next, collect the insoluble matter on a tared Gooch crucible, wash with hot water, ignite, and weigh. Determine the percent of acid-insoluble ash calculated from the weight of the leaves used. Record results:

$$\frac{\text{Weight of ash in crucible}}{\text{Weight of leaves}} \times 100 \underline{\hspace{1cm}} \%$$

## SELECTED REFERENCES—GLYCOSIDES

Britto-Arias, M. *Synthesis and Characterization of Glycosides*. Springer, 2007.
Goodwin, T. *Introduction to Plant Biochemistry*. Pergamon, 1983.
Lindhurst, T. *Essentials of Carbohydrate Chemistry and Biochemistry*. Wiley, 2007.
Levy, D. *The Chemistry of Glycosides*. Pergamon, 1995.
Miller, L. P. *Phytochemistry* (3 vol.). Van Nostrand, 1973.
Murray, R. D. H. *The Natural Coumarins*. Wiley, 1989.

Ortega, M. *New Developments in Medicinal Chemistry*. Nova Science, 2009.
Sim, S. K. *Medicinal Plant Glycosides*. Toronto University Press, 1967.
Thompson, R. H. *Naturally Occurring Quinones*. Academic, 1971.
Thompson, R. *The Chemistry of Natural Products*. Blackie, 1985.
Yang, C. R. *Advances in Plant Chemistry*. Elsevier, 2012.

# 14 Gums

## GUMS

Gums are natural plant hydrocolloids that may be classified as anionic or nonionic polysaccharides or salts of polysaccharides. Chemically, the gums, along with mucilages, pectins, and celluloses, are condensations of pentoses and/or hexoses. Gums and mucilages are closely related to the hemicelluloses in composition and form.

Some common gums used in medicinal products include tragacanth, arabic (acacia), sterculic (karaya), agar, and plantago seed. The latter of these is used as a cathartic whose action is due to the swelling of the mucilaginous seed coat, thus giving both bulk and lubrication. Agar has an algae plant base and has been used as a bulk laxative as well as a bacteriological culture medium. Gum karaya is also used as a bulk laxative while tragacanth is employed in hand lotions or as a demulcent and emollient and acacia as a suspending agent.

Experiments in this section include an investigation of some physical and chemical features of different gums and illustrations of the mucilage-swelling factor in gums.

## EXPERIMENT 53: SOME CHEMICAL FEATURES OF GUMS

Gums are natural hydrocolloids that may be classified as anionic or nonionic polysaccharides or salts of polysaccharides. They may therefore be categorized under the heading of carbohydrates, but due to their special importance to biological and pharmaceutical chemistry, are being considered in their own special section.

Gums are amorphous and translucent substances frequently produced in higher plants as protectives after an injury. They are typically heterogeneous in composition and upon hydrolysis yield arabinose, galactose, glucose, mannose, xylose, and various uronic acids.

Gums find diverse applications in medicine and pharmacy, including dental adhesives, bulk laxatives, tablet binders, emulsifiers, gelating agents, stabilizers, and thickeners.

Gums, or hydrocolloids, can be summarized as follows:

1. Shrub or tree exudates—acacia, karaya, tragacanth
2. Marine gums—agar, algin, carrageenin
3. Seed gums—guar, locust bean, psyllium
4. Plant extracts—pectins
5. Starch and cellulose derivatives—ethyl and methyl cellulose
6. Microbial gums—dextran, xanthan

The varying stability of gums in different solvents is an important consideration in pharmaceutical compounding. The first activity in this section will be designed to measure the relative solubilities of four gums in solutions of water, alcohol, and acid.

To begin the experiment, set up 12 125 mL Erlenmeyer flasks labeled 1 through 12, as shown in the illustration.

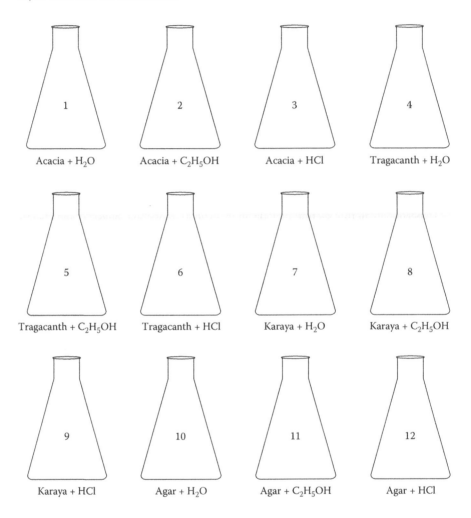

| 1 | 2 | 3 | 4 |
| Acacia + H$_2$O | Acacia + C$_2$H$_5$OH | Acacia + HCl | Tragacanth + H$_2$O |

| 5 | 6 | 7 | 8 |
| Tragacanth + C$_2$H$_5$OH | Tragacanth + HCl | Karaya + H$_2$O | Karaya + C$_2$H$_5$OH |

| 9 | 10 | 11 | 12 |
| Karaya + HCl | Agar + H$_2$O | Agar + C$_2$H$_5$OH | Agar + HCl |

In flasks 1, 4, 7, and 10, place 25 mL of water. In flasks 2, 5, 8, and 11, place 25 mL of 95% C$_2$HH$_5$OH. In flasks 3, 6, 9, and 12, place 25 mL of 1 N HCl.

The next step is to measure out 2 g of acacia and add to flask 1. Repeat this procedure for flasks 2, 3. In flasks 4, 5, and 6 add 2 g of gum tragacanth; in flasks 7, 8, 9, and add 92 g of gum karaya; and in flasks 10, 11, 12, and add 122 g of agar. When you have completed this, it will mean that each of the four gums will be tested with each of the three solvents.

You will need to shake each of the 12 flasks vigorously to achieve as much solubility as possible.

The next step is to set up a filtration funnel rack using previously weighed sections of cheesecloth to filter all the fluid mixtures. Record all original weights of the pieces of cheesecloth in the data. Once filtration is complete, place the filters in a drying oven at 103°C for 1 hour. Remove from the oven after the hour has passed, cool, and then weigh each filter and residue. Record data.

Calculate the weight of soluble material in each gum sample by subtracting the original weight of the filter from the final weight of the filter and residue. Record data and calculations.

---

**Model for Calculations:**

| | | |
|---|---|---|
| Filter + residue weight | = | _____ g |
| – Original filter weight | = | _____ g |
| Weight of insoluble material | = | _____ g |

---

$$\text{Percent of insoluble gum material} = \frac{\text{wt of insolubles}}{\text{original wt (2 g)}}$$

$\times\ 100 =$ percent of insoluble material in each of the 12 solvent systems.

$100 -$ insoluble percent = percent of soluble material

Follow this pattern for calculations on all 12 systems.

Results: _____.

On the basis of your calculations, which of the following gums is:

The most water soluble? _____
The least water soluble? _____
The most alcohol soluble? _____
The least alcohol soluble? _____
The most acid soluble? _____
The least acid soluble? _____

## EXPERIMENT 54: THE MUCILAGE-SWELLING FACTOR IN GUMS

The mucilage-swelling factor is an important physical characteristic of gums, which is useful in medicinal preparations. To test this feature take four 50 mL beakers and place 2 g of plantago (psyllium) seed in the first, 2 g of carrageenin (Irish moss) in the second, 2 g of locust bean (carob) in the third, and 2 g of guar gum in the fourth. Now measure out 40 mL of water in a small graduated cylinder and add it to the first beaker. Repeat this procedure for the remaining three beakers. Next place a stirring magnet in each beaker and place each over a magnetic stirrer and run for 12 hours. At the end of this period, note the mucilage level of each beaker and record results.

1. Plantago _____ mL
2. Carrageenin _____ mL
3. Locust bean _____ mL
4. Guar gum _____ mL

Check some references in pharmaceutical chemistry and explain how the mucilage-swelling factor may be a consideration for the compounding of various medicinals.

How may differences in solubilities in various solvent systems affect the selection of gums in various preparations?

## SELECTED REFERENCES—GUMS

Atkinson, E. *Gums and Gum Resins*. Bibliolife, 2010.

Davidson, R. L. *Handbook of Water-Soluble Gums*. McGraw-Hill, 1989.

Dieterich, K. *Analysis of Resins, Balsams, and Gums*. Book Depository, 2001.

Gokhale, S. B. *Practical Pharmacognosy*. Prakashan, 2008.

Hoppe, H. A. *Marine Algae in the Pharmaceutical Sciences*. Guyter, 1980.

Perry, E. *Gums*. General Books, 2010.

Salter, C. *Analysis of Resins, Balsams and Gum Resins*. Biblilife, 2010.

Smith, E. L. *Principles of Biochemistry*. McGraw-Hill, 1983.

Whistlen, R. L. *Industrial Gums*. Academic Press, 1973.

Youngken, H. *Textbook of Pharmacognosy*. Blakiston, 1950.

# 15 Balsams

## BALSAMS

Balsams are resinous mixtures that contain large proportions of benzoic or cinnamic acid, or both, or esters of these acids. The medicinally-sed balsams include Peru balsam. Tolu balsam, styrax, and benzoin. Benzoic acid may be readily sublimed from balsam of Peru, balsam of tolu, and benzoin. Cinnamic acid is readily sublimed from balsam of Peru, balsam of tolu, styrax, and Sumatra benzoin.

Pharmaceutical use of balsams include the use of Peru balsam as a parisitide in skin diseases. Tolu balsam is a pharmaceutic necessity used in the compound tincture of benzoin. It is also used as an expectorant and as a flavoring in medicinal syrups, confectionery, chewing gum, and perfumery. Benzoin, in the form of benzoic acid, is sometimes used as a mild antiseptic, stimulant, and expectorant, and externally as a wound dressing. It also occasionally finds use in veterinary medicine as an expectorant and antipyretic.

Experiments in this section include an investigation of acid value of balsams and some tests which serve to illustrate the differences between Siam (Thailand) and Sumatra Benzoin.

## EXPERIMENT 55: DETERMINING THE ACID VALUE OF BALSAMS

Balsams are resinous mixtures that contain large proportions of benzoic or cinnamic acid or both, or esters of these acids. The medicinal balsams include Peru balsam, tolu balsam, styrax, and benzoin.

Peru balsam, also called balsam of Peru, appeared in the U.S.P. from 1820 to 1960. It is obtained from trees abundant along the coast of San Salvador in Central America. It was imported to Spain by early explorers via Lima, Peru, hence the name. It is a pathological product being formed by injury to the tree that causes the balsam to exude from the exposed wood.

Tolu balsam is obtained from trees growing abundantly along the lower Magdalena River in Colombia but is also found in Venezuela and the West Indies. Like Peru balsam, it is also obtained by making incisions through the bark and sap wood.

Styrax, or storax, is a balsam obtained from both oriental (Levant) and North American sources, the latter sometimes being called sweet gum. It has a very ancient history, being mentioned by Arab physicians in the 12th century. Most of the styrax used in pharmacy comes from Turkey.

A discussion of benzoin appears in the next experiment. In this experiment, we will perform titrations to determine the acid values of Peru balsam, tolu balsam, Levant storax, and American storax by following the same procedure with samples of each of the four substances and recording the data.

Dissolve 1 g of the substance being tested in 50 mL of methanol and add 0.5 mL of phenolphthalein t.s. Next titrate using buret and magnetic stirrer with 0.5 N NaOH and record the acid value.

|               | Acid Value |
|---------------|------------|
| Peru balsam   | _____ |
| Tolu balsam   | _____ |
| Levant storax | _____ |
| American storax | _____ |

The final step in this investigation is to check acid value parameters given in the USP/NF or pharmacognosy texts to determine the accuracy of your results.

Check pharmacy references and list some commercial preparations that contain various balsams.

## EXPERIMENT 56: DIFFERENTIATION OF SIAM AND SUMATRA BENZOIN

Benzoin is a balsamic resin obtained from two major sources, which gives rise to its name, either Siam or Sumatra benzoin. Siam benzoin is obtained from trees growing in Thailand (formerly called Siam), and in Annam and Tonkin, whereas Sumatra benzoin comes from trees native to southeastern Asia and the East Indies. They first appeared in the literature when mentioned by an Arab pharmacist in the 14th century, and by the 16th century it had become an object of Venetian commerce.

The use of Siam benzoin is confined almost entirely to the perfumery industry, whereas Sumatra benzoin finds much wider use in pharmacy.

The object of this experiment is to perform two tests that will provide ways of distinguishing the two products.

Dissolve 5 mL of Siam benzoin in 20 mL of ether and pour the solution into a porcelain dish. Add three drops of concentrated $H_2SO_4$ and record the results. Now repeat the procedure, this time using Sumatra benzoin and again record results.

Siam benzoin color reaction _____
Sumatra benzoin color reaction _____

The second set of tests to distinguish between the two benzoins is to warm 500 mg of powdered Siam benzoin with 10 mL of $KMnO_4$ t.s. and gently warm. Record any noticeable odor. Repeat the procedure using Sumatra benzoin and record any evidence of odor.

Siam benzoin _____
Sumatra benzoin _____

The next step is to check the accuracy of your recorded results using the USP/NF or texts in pharmacognosy or pharmaceutical chemistry.

A final activity in this experiment is to compile some of the uses of benzoin in commercial pharmacy using *Remington's* and other pharmacy reference texts. Record information below.

## SELECTED REFERENCES—BALSAMS

DeMayo, F. *The Chemistry of Natural Products.* Interscience, 1959.
Dieterich, K. *Analysis of Resins, Balsams, and Gums.* Book Depository, 2001.
Gokhale, S. B. *Practical Pharmacognosy.* Prakashan, 2008.
Howes, F. N. *Vegetable Gums and Resins.* Chronical Botanica, 1949.
Rahman, A. *Studies in Natural Products.* Elsevier, 2012.
Robertson, P. *Pharmacognosy: The Study of Natural Drug Substances.* Lippincott, 1956.
Salter, C. *Analysis of Resins, Balsams, and Gum Resins.* Bibliolife, 2010.
Youngken, H. *Textbook of Pharmacognosy.* Blakiston, 1950.

# 16 Volatile Oils

## VOLATILE OILS

Volatile oils are widespread in the plant family and may occur in specialized secretory structures, in modified parenchyma cells, in oil tubes called *vitae*, or in lysigenous or schizogenous passages. Volatile oils may act as repellants to insects, thus preventing the destruction of flowers and leaves, or they may serve as insect attractants, thus aiding in the cross-fertilization of the flowers.

On the basis of their biosynthetic origin volatile oils may be divided into two broad categories:

1. Terpene derivatives formed via the acetate–mevalonic acid pathway or
2. Aromatic compounds formed via the shikimic acid and phenylpropanoid route.

Modern medicine makes use of such volatile oils as peppermint oil, which continues to find use as a carminative and a flavor; menthol, employed as a local antipruritic as well as counter irritant, and pine oil, used as a disinfectant especially in the veterinary field.

Experiments in this section involve steam distillation of some volatile oils from plant sources as well as assays of several volatile oil products to determine their chemical constituents.

## EXPERIMENT 57: STEAM DISTILLATION OF PEPPERMINT OIL

Volatile oils are the odorous principles found in various plant parts. Because they evaporate when exposed to air at ordinary temperatures, they are called volatile or essential oils. They are generally immiscible with water, but are sufficiently soluble to impart their odor to water. They are usually soluble, however, in ether, alcohols, and other organic solvents.

Many volatile oils are used in medicinal preparations. These include peppermint, menthol, cardamom, rose, juniper, pine, orange peel, cedarwood, camphor, spearmint, clove, thyme, birch, anise, sassafras, eucalyptus, and others.

Steam distillation is a means of separating and purifying organic compounds by volatilization. In this experiment, we will separate peppermint oil from its leaves using this method. Most compounds, regardless of their natural boiling point, will distill by steam distillation at a temperature below that of the pure boiling point. This has become a very effective means of separating volatile oils from natural sources, since it involves less heat and hence less risk of destroying the compound you seek to isolate.

Place 3 g of peppermint leaves in a 500 mL flask containing 200 mL of water. Connect this to a steam generator as shown in the figure. The flask with the peppermint leaves and water should then be connected to a condenser and a receiving flask for the volatile oil. *Caution:* Make sure the safety pressure tube in the steam generator is below the surface of the water at all times to prevent excessive pressure and a possible explosion. Begin heating the water in the steam generator until the volatile condensate begins to appear in the receiving flask. Continue this process until no more condensate appears in the receiving flask. Discontinue heating, cool down the apparatus, and measure the milliliter of peppermint oil collected.

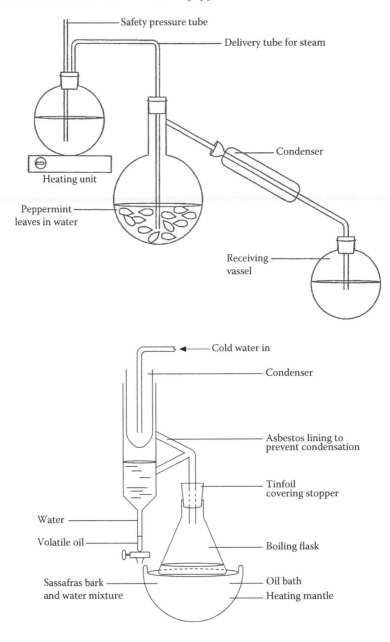

To conclude this section on volatile oils, check *Remington's*, the USP/NF, and other pharmaceutical references and make a list of the various volatile oils that are used in medicinal preparations today.

Preparations containing volatile oils:

## EXPERIMENT 58: ASSAY OF BITTER ALMOND OIL FOR HYDROCYANIC ACID

Bitter almond oil is about 80% benzaldehyde and 2%–4% hydrocyanic. The type used for food flavorings has the HCN removed, but the other nonadulterated type is still used as a perfume for lotions.

Procedure for the assay: Dissolve 750 mg of $MgSO_4$ in 45 mL of distilled water and add 5 mL of a 0.5 M NaOH solution and 2 drops of $K_2CrO_4$ t.s. Titrate the solution with 0.1 M $AgNO_3$ to the production of a permanent reddish color.

The next step is to pour this mixture into a 100 mL flask containing 1 g of bitter almond oil and titrate again using a magnetic stirrer with 0.1 M $AgNO_3$ until a red color comes, which does not disappear. Each milliliter of 0.1 $AgNO_3$ is equivalent to 2.703 mg of hydrocyanic acid (HCN).

The reactions for this process are

1. $MgSO_4 + 2\ NaOH \rightarrow Mg(OH)_2 + Na_2SO_4$

2. 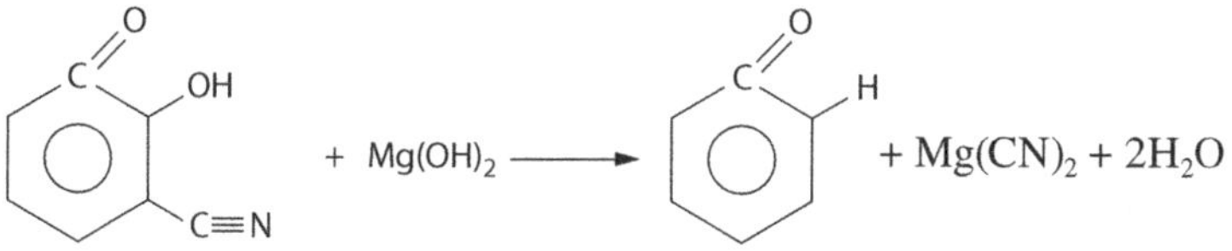 $+\ Mg(OH)_2 \longrightarrow$ $+\ Mg(CN)_2 + 2H_2O$

3. $Mg(CN)_2 + 2\ AgNO_3 \rightarrow AgCN + Mg(NO_3)_2$
4. $K_2CrO_4 + 2\ AgNO_3 \rightarrow Ag_2SO_4 + 2\ KNO_3$

Record results:

_____ mL (0.1M $AgNO_3$) × 2.703 _____ HCN

## EXPERIMENT 59: ANALYSIS OF THUJONE IN ARBOR VITAE OIL

The arbor vitae shrub, known botanically as *Thuja occidentalis*, is native to North America. The oil is obtained from the plant's leafy young twigs by steam distillation. It has been used in pharmacy to treat warts and as a counterirritant. The formula is

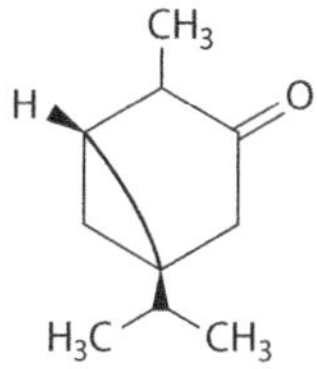

Molecular formula is $C_{10}H_{16}O$. Mole weight is 152.23. *Procedure*: Dissolve 5 g of hydroxylamine HCl in 9 mL of warm water, add 80 mL of 90% ethanol and 2 mL of bromophenol blue t.s. Neutralize this mixture with 0.5 M KOH (alcoholic) and add sufficient ethanol to make 100 mL. Then weigh 1 g of thuja oil and 20 mL of the hydroxylamine HCl. Shake the mixture and titrate it with 0.5 M KOH (alcoholic). Continue titration for 4 hours at intervals of 30 minutes.

Each milliliter of the 0.5 M KOH (alcoholic) is equivalent to 0.0761 g of ketone calculated as Thujone.

Record results:

_____ mL KOH × 0.0761 _____ g of Thujone

## SELECTED REFERENCES—VOLATILE OILS

Brauerman, J. B. *Citrus Products*. Interscience, 1949.
DeMayo, P. *The Chemistry of Natural Products*. Interscience, 1959.
Guenther, E. *The Essential Oils* (6 vol.). Van Nostrand, 1959.
Lawless, J. *The Illustrated Encyclopedia of Essential Oils*. Element Publishing, 1995.
Lawless, J. *The Encyclopedia of Essential Oils*. Thorson Publishing, 2002.
Shugar, G. *Chemistry Technician's Ready Reference*. McGraw-Hill, 1990.
Simonsen, J. *The Terpenes*. Cambridge University Press, 1959
Stewart, D. *Chemistry of Essential Oils Made Simple Life*. Science Publications, 2009.
Williams, D. *The Essential Oils*. Lavandes, 2006.
Young, D. *Essential Oils*. Product Reference Life, 2011.

# 17 Analgesics

## ANALGESICS

Analgesics may be defined as drugs that bring about insensibility to pain without loss of consciousness. Their use started with plant drugs that were discovered to relieve pain and broadened to the modern development of synthesizing molecules of effective agents. The discovery of morphine in the early 19th century launched a search for other effective plant drugs.

Alternatives to the morphine molecule have led to the development of many effective analgesics, including codeine, hydrocodone, meperidine, oxycodone, and methadone, to name but a few. In addition to this, there are also antipyretic analgesics, which include acetylsalicylic acid, acetaminophen, phenylbutazone, and naproxen.

Experiments in this section involve analytical methods for identifying various analgesics plus the preparation of several analgesics that can be synthesized in the laboratory,

## EXPERIMENT 60: IDENTIFICATION OF ANALGESICS USING TLC

There are four basic categories of analgesics on the over-the-counter drug market. OTC pain killers have either acetylsalicylic acid, acetaminophen, ibuprofen, or naproxen as their main active ingredient. Formulas for the four compounds are

Acetylsalicylic acid       Acetaminophen

Ibuprofen       Naproxen

In this experiment, you will use two commercially prepared TLC plates with silica gel backing. The first plate will be spotted by solutions of the four substances listed above and the second plate will be spotted with solutions of four unknown OTC commercial analgesic preparations.

## TLC Procedure

The two TLC plates of approximately $10 \times 6$ cm should be handled only at the edges so as not to interfere with later test results. Using a pencil (but not a pen), lightly draw a line across the plates (short dimension) about 1 cm from the bottom. Lightly mark four spots on each plate to identify where you will introduce your spots.

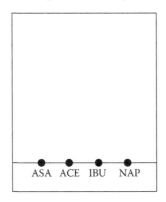
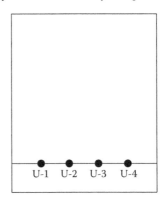

Prepare, or have prepared ahead of time, eight solutions, containing 1 g of each of the four analgesics, plus the four unknowns dissolved in 20 mL of a 50–50 mixture of methylene chloride and ethanol. These solutions will be spotted on the plates using micropipets.

    As each plate spotting is completed, place them into a suitable development chamber such as a wide-mouth screw-cap jar lined with filter paper soaked with the development solvent. In this case, the solvent consists of 0.5% glacial acetic acid in ethyl acetate. This liner is designed to keep the chamber saturated with solvent vapors, thus facilitating the process. On the bottom of the development chamber prior to introducing the plates, pour the acetic acid–ethyl acetate solvent mixture to a depth of just under 1 cm. Measure this carefully to ensure the solvent level in the chamber is not above the spot that was applied to the plate; otherwise, the spotted material will dissolve in the pool of the solvent instead of undergoing chromatography. The setup should look like the following figure:

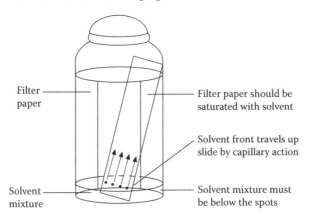

Movement of the solvent front occurs quite rapidly. The plate should be removed from the development chamber when the front is within about 1 cm of the top and should not be allowed to run to the top edge. Set the plates on a paper towel to dry once they have been removed from the chamber. When they are dry, observe them under UV illumination in a darkened room and observe placing and coloration of the spots. Mark with a pencil so you can calculate and record $R_f$ values for the eight substances.

An additional test consists of placing a plate in a small jar containing iodine crystals. Cap the jar and warm gently on a steam bath until the spots begin to appear. Notice which spots become visible and note their relative colors. You are now ready to draw some conclusions on the unknowns as you compare these results with those of the known analgesics.

| Substance | $R_f$ Value | $I_2$ Reaction |
|---|---|---|
| Acetylsalicylic acid | _____ | _____ |
| Acetaminophen | _____ | _____ |
| Ibuprofen | _____ | _____ |
| Naproxen | _____ | _____ |
| U-1(_____) | _____ | _____ |
| U-2(_____) | _____ | _____ |
| U-3(_____) | _____ | _____ |
| U-4(_____) | _____ | _____ |

## EXPERIMENT 61: SYNTHESIS OF ACETYLSALICYLIC ACID

Commonly called *aspirin*, and an ingredient in many analgesic remedies, acetylsalicylic acid can easily be synthesized in the laboratory and a percentage yield calculated. The equation for the reaction, which we will carry out is

Salicylic acid          Acetic anhydride

Acetylsalicylic acid          Acetic acid

### PROCEDURE FOR OBTAINING ASA

Place 6 g of salicylic acid and 10 mL of acetic anhydride in a 500 mL Erlenmeyer flask. Cautiously add 10 drops of concentrated $H_2SO_4$ and swirl gently. Now warm the solution until all the solid is dissolved. Remove the flask from the water bath and allow a few minutes for cooling. Then place the flask in a large beaker containing ice cubes and wait about 15 minutes until complete crystallization occurs. Weigh a piece of filter paper and record on your data sheet, then spoon off the solid from the flask to the filter paper, which is fitted to a Büchner funnel held by a rubber stopper on a vacuum flask and connected either to an aspirator on the lab faucet or a vacuum pump. Suction the crystals to as dry a condition as possible and then place in a drying oven at a temperature of 103°C for a period of an hour.

After the paper with the powder has been removed from the oven, cool it for 10 minutes, record results, and calculate the percentage yield based on the equation for the reaction:

Weight of filter paper and product _____ g
Weight of filter paper _____ g
Weight of ASA obtained _____ g

The percentage yield can then be calculated:

$$\text{Percentage yield } \frac{\text{Actual weight} \quad \text{g}}{\text{Theoretical weight g}} \times 100 = \_\_\_\_\_ \%$$

## EXPERIMENT 62: THE PREPARATION OF ACETAMINOPHEN

Acetaminophen is one of the four major categories of commercial analgesics and can be prepared from purified p-aminophenol according to the formula:

p-aminophenol     Acetic anhydride          Acetaminophen          Acetic acid

Weigh out about 0.1 g of purified p-aminophenol and place it in a 3 mL conical vial at the bottom of which is a small magnetic spinner. Using an automatic or graduated pipet, add 0.3 mL of water and 0.11 mL of acetic anhydride. Attach the vial to a small air condenser and heat the reaction in a sand bath at about 115°C. Continue heating until the solid has completely dissolved in the acetic anhydride. Allow the vial to cool while scratching its side with a glass rod to induce crystallization. This process results in the formation of crude acetaminophen.

In order to purify your product, you can use solvent mixtures consisting of a 50:50 mixture of water and methanol by volume. Crystallization may be done using either a Craig tube or a Hirsch funnel. Either have the advantage over simple precipitation reactions since in the process of crystallization, as a rule, the crystal growth tends to be relatively slow and selective, whereas in precipitation the process is rapid and nonselective, hence allowing more opportunity for the trapping of impurities.

Once you have recrystallized a purified product, use stoichiometric computations to calculate a percentage yield.

Record:

$$\frac{\text{Actual weight product} \quad g}{\text{Theoretical weight of product} \ g} \times 100 = \underline{\hspace{1cm}} \%$$

## SELECTED REFERENCES—ANALGESICS

Barlow, R. B. *Introduction to Chemical Pharmacology*. Wiley, 1955.

Burger, A. *Medicinal Chemistry*. Interscience, 1960.

Black, J. *Wilson and Gisvold's Organic, Medicinal, and Pharmaceutical Chemistry*. Lippincott, 2011.

Doerge, R. *Wilson and Gisvold's Organic, Medicinal, and Pharmaceutical Chemistry*. Lippincott, 1982.

Fellows, E. J. *Analgesics*. Wiley, 1951.

Janesn, F. *Synthetic Analgesics*. Pergamon Press, 1960.

Lemke, T. *Foye's Principles of Medicinal Chemistry*. Lippincott, 2010.

Ortega, M. *New Developments in Medicinal Chemistry*. Nova Science, 2009.

Patrick, G. *An Introduction to Medicinal Chemistry*. Oxford University. Press, 2009.

Watson, D. *Pharmaceutical Analysis*. Elsevier, 2012.

Wilson, C. O. *Textbook of Organic, Medicinal, and Pharmaceutical Chemistry*. Lippincott, 1962.

# 18 Anesthetics

## ANESTHETICS

Anesthetics are drugs that produce a loss of sensation and or motor function by causing a block of nerve conduction. Ideally, they are nonirritating to tissue, have low systemic toxicity, are effective whether injected or applied locally, and have a rapid onset of anesthesis, but short duration of action. One of the earliest uses was that of cocaine in 1860.

Cocaine is a benzoic acid derivative. There are also anesthetics that are p-aminobenzoic acid derivatives such as procaine, benzocaine, and anilide derivatives, including lidocaine. In addition to general anesthetics, there are also numerous local anesthetics such as dibucaine ointments and benzocaine cream, plus topical anesthetics such as tetracaine solution and ophthalmic solutions such benoxinate hydrochloride.

Experiments in this section will focus on laboratory preparation of the anesthetic benzocaine, with calculation of percentage yield. The second experiment is designed to illustrate how different molecular structures of anesthetics can be identified by analytical methods.

## EXPERIMENT 63: THE PREPARATION OF BENZOCAINE

The two categories of anesthetics are general and local. General anesthetics may be either volatile substances administered by inhalation or nonvolatile drugs administered by routes other than inhalation. Inhalation anesthetics include nitrous oxide, ethylene, and cyclopropane. Nonvolatile anesthetics suitable for routes other than inhalation include barbiturates and the various narcotics, which produce, in addition to anesthesia, analgesic, and hypnotic effects. Local anesthetics have such subcategories as epidural, regional, nerve blockers, spinal, and topical. Included here are such compounds as cocaine, novocaine, and benzocaine.

Most chemically useful anesthetics have common core structures, which consist either of amino compounds or nonnitrogenous hydroxy compounds. The amino compounds have a cyclic structure, usually aromatic, and an amino-nitrogen group separated by an intervening aliphatic chain consisting of two or three carbon atoms:

$$R_3 - \bigcirc - CH_2\ CH_2\ N \begin{array}{c} R_1 \\ R_2 \end{array}$$

## PREPARING BENZOCAINE

Benzocaine is a local anesthetic produced by the direct esterification of p-aminobenzoic acid with ethanol. The equation for the reaction is

p-aminobenzoic acid                    Ethyl p-aminobenzoate
                                       (benzocaine)

Place 2.4 g of p-aminobenzoic acid and 12 mL of 95% ethanol into a 250 mL boiling flask. Add a magnetic stirrer and stir until the solid dissolves completely. With stirring, add 2 mL of concentrated $H_2SO_4$, which will result in the formation of a large precipitate. Attach this to a water-cooled condenser and begin refluxing with a water bath for about 1 hour at 115°C.

Once this process is complete, remove the apparatus from the water bath and allow the reaction to set for several minutes. Then add 10 mL of a 10% $Na_2CO_3$ solution, which will result in the evolution of considerable gas (frothing) until the solution is neutralized. When the gas has completed its evolution add sufficient additional $Na_2CO_3$ solution to bring the mixture to a pH of 8.0. This should result in the formation of a white precipitate of benzocaine.

You are now ready to collect the benzocaine using a Büchner funnel and either an aspirator or vacuum arrangement. It will be helpful to wash the solid thoroughly with water. Record the weight of the fair pure benzocaine once it has been air-dried overnight. Place a few milligram of the benzocaine in the melting point apparatus and determine the melting point. Check with information in the Merck Index to determine how close your results are to the official data.

Weight of benzocaine obtained _____ g
Theoretical weight _____ g
Melting point of your product _____ °C
Melting point (official) _____ °C

Check the pharmaceutical literature, local drug stores, and the lab stockroom, and list various analgesic preparations that contain benzocaine:

_____
_____
_____
_____
_____

## EXPERIMENT 64: TWO CATEGORIES OF GENERAL ANESTHETICS

In contrast to benzocaine, which is a topical anesthetic, lidocaine (trade name Xylocaine) and procaine (trade name Novocaine) are examples of drugs that possess general anesthetic properties, although having different molecular structures. Whereas lidocaine is chemically 2-diethyl-2,6 acetoxylidide, procaine is a derivative of para-amino benzoic acid, which chemically is $p$-amino-$N$-[2-(diethylamino)ethyl] benzamide hydrochloride. The respective formulas are

(Lidocaine)

(Procaine)

In this experiment, comparisons will be made that distinguish the two categories of molecular structures. Begin by placing 100 mg of lidocaine dissolved in 1 mL of $CH_3OH$ and adding 10 drops of $CoCl_2$ t.s. Shake for 2 minutes and record color. Repeat this procedure using procaine HCl and again record the color reaction.

Lidocaine with $CoCl_2$ _____.
Procaine with $CoCl_2$ _____.

The next procedure involves taking 100 mg of lidocaine and dissolving it in a mixture of 5 mL water and 1 mL of dilute $HNO_3$. Then add 3 mL of $Hg(NO_3)_2$ and heat to boiling. Record color. Repeat this procedure using procaine HCl and record results.

Lidocaine with $Hg(NO_3)_2$ _____.
Procaine with $Hg(NO_3)_2$ _____.

There are, of course, many other chemical structures that possess either general or local anesthesia on the body, including hydrocarbon derivatives such as ethyl chloride and halothane and other barbituric acid structures such as hexobarbital sodium.

Develop a list of some major general and local anesthetics that are used extensively in medicine dentistry. Consult references.

## SELECTED REFERENCES—ANESTHETICS

Adriani, J. *Chemistry and Physics of Anesthesia Better.* World Books, 2010.
Black, J. *Wilson and Gisvold's Organic, Medicinal, and Pharmaceutical Chemistry.* Lippincott, 2011.

Burger, A. *Medicinal Chemistry*. Interscience, 1960.
Lawton, G. *Progress in Medicinal Chemistry*. Elsevier, 2011.
Lemke, T. *Foye's Principles of Medicinal Chemistry*. Lippincott, 2010.
Lofgren, A. *Studies on Local Anesthetics*. University of Stockholm, 1948.
Patrick, G. *An Introduction to Medicinal Chemistry*. Oxford University Press, 2009.
Pavia, D. *Organic Laboratory Techniques*. Saunders, 1988.
Wilson, C. O. *Textbook of Organic, Medicinal, and Pharmaceutical Chemistry*. Lippincott, 1962.

# 19 Sulfa Drugs (Sulfonamides)

## SULFA DRUGS

Sulfonamides came on the drug scene in the 1930s when azo dyes research indicated their antibacterial effectiveness. While more than 4,000 were tested, only about a dozen found use in therapeutic medicine. The sulfonamides include

1. Aniline—substituted ones such as the sulfanilimides
2. Prodrugs such as Sulfasalazine
3. Nonaniline ones such as Mafenide

Sulfonamides are used less now than they were in the past since many bacterial strains have developed an effective resistance against them, although a few are still used in topical ointments and ophthalmic solutions.

Experiments in this section involve the preparation of sulfanilamide, differential reactions of different sulfonamides, assays and analysis of some sulfa drugs, plus a chromatographic plate identifying sulfonamides.

## EXPERIMENT 65: THE PREPARATION OF SULFANILAMIDE

The basic sulfanilamide molecule was the first of the so-called sulfa drugs. It is made by converting acetanilide to *p*-acetamidobenzenesulfonyl chloride and from there to sulfanilamide according to the following equation:

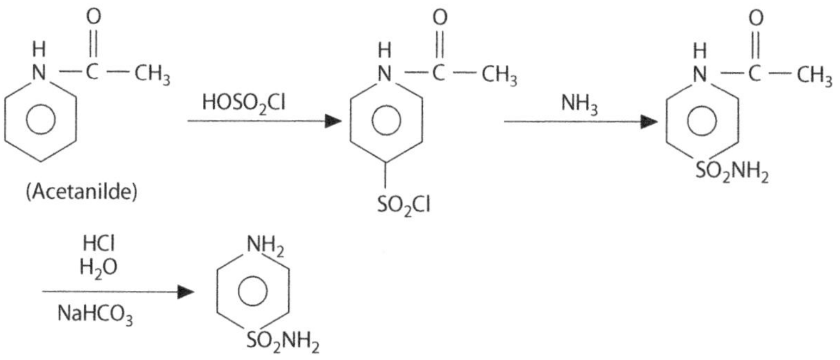

(Acetanilde)

Sulfanilimide

*Procedure*: Assemble an apparatus as pictured in the following illustration using a 5 mL conical vial and an air condenser placed in a sand bath at about 120°C.

Place 0.18 g of acetanilide in the conical vial and connect the air condenser. Melt the acetanilide (mp 113°C) by heating the vial in the sand bath. Remove the vial from the sand bath and remove the heavy oil while holding the vial so that it solidifies uniformly on the cone-shaped bottom of the vial. Allow the vial to cool to room temperature and then cool it further in an ice-water bath.

Remove the air condenser. In a hood, transfer 0.5 mL of chlorosulfonic acid (*caution*: very corrosive) to the acetanilide in the vial using a graduated pipet. Reattach the air condenser. Allow the mixture to stand for several minutes, then heat the vial in the sand bath to about 120°C for 10 minutes to complete the reaction. Now add 3 g of crushed ice to a 20 mL beaker, and in a hood, transfer the cooled reaction mixture dropwise, using a Pasteur pipet. Rinse the conical vial with a few drops of cold water and transfer the contents to the beaker containing ice. Stir the precipitate to break up the lumps and then filter the *p*-acetamidobenzenesulfonyl chloride to a Hirsch funnel. Rinse the conical vial and beaker with two 1 mL portions of ice water.

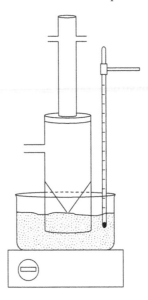

Prepare a hot water bath at 70°C. Place the crude product into the original 5 mL conical vial and add 1.1 mL of dilute NH$_4$OH (50:50 concentrated NH$_4$OH and H$_2$O). Stir the mixture well with a spatula and reattach the air condenser. Heat the mixture in the hot water bath for 10 minutes. Allow the vial to cool and then place in the ice bath for several minutes. Collect the *p*-acetamidobenzenesulfonamide in a Hirsch funnel and rinse the vial with water.

Transfer this solid into the vial and add 0.53 mL of dilute HCl (3.6:7.0 mL of concentrated HCl:H$_2$O). Attach the air condenser and heat the mixture in the sand bath at about 140°C until all the solids have dissolved. When the vial has cooled to room temperature, no further solids should appear. With a Pasteur pipet, transfer the solution to a 20 mL beaker. While stirring with a glass rod, cautiously add dropwise a slurry of 0.5 g of NaHCO$_3$ in about 1 mL of H$_2$O to the mixture in the beaker. Foaming will occur after each addition of the NaHCO$_3$ due to the evolution of $CO_2$ gas. Allow the evolution to cease. Eventually, sulfanilamide will begin to precipitate. At that point, begin to check the pH of the solution and add aqueous $NaHCO_3$ until the pH is between 5.0 and 6.0. Cool the mixture thoroughly in an ice water bath. Collect the product on a Hirsch funnel and rinse the beaker and solid with about 0.5 mL of cold water. Allow the sulfanilamide to air dry. In the next laboratory period, weigh the dry product and calculate the percentage yield.

$$\frac{\text{Weight of acetanilide} - 0.18\,\text{g}}{\text{Acetanilide m.w.} - 180} = \frac{\text{Theoretical wt (sulfanilirnide} \times \text{g})}{\text{Sulfanilirnide m.w.} - 172}$$

$$\text{Percentage yield} = \frac{\text{Actual weight of sulfanilimide}}{\text{Theoretical weight of sulfanilirnide}} \times 100 = \underline{\hspace{1cm}} \%\,\text{yield.}$$

## EXPERIMENT 66: REACTIONS OF SELECTED SULFONAMIDES

The current *Merck Index* lists more than 70 sulfonamides, most of which possess antibacterial properties, with varying physiological side effects. The original so-called sulfa drug was sulfanilimide, but newer drugs with similar effects and less toxicity have replaced its use in pharmacy. In this section, we will study six of them as samples of the sulfonamide category:

Sulfadiazine

Sulfathiazole

Sulfamethazine

Sulfaacetamide

Sulfamerazine

Sulfapyridine

1. Gently heat 1 g of sulfadiazine in a small test tube until its sublimate is formed. Collect a few milligram of the sublimate with a glass rod and mix it in another test tube with 1 mL of L:20 resorcinol:ethyl alcohol. Add 1 mL of $H_2SO_4$, mix, and shake. Record the color reaction _____.
2. To 100 mg of sulfathiazole, add 5 mL of 3 N HCl and boil it gently for 5 minutes. Cool in an ice water bath and add 4 mL of $NaNO_2$ solution (1%) and dilute with $H_2O$ to 10 mL. Place the mixture in an ice bath for 10 minutes. To 5 mL of the cooled mixture, add a solution of 50 mg betanaphthol in 2 mL of 10% NaOH and record the color reaction _____.
3. Dissolve 100 mg of sulfacetamide in 5 mL of $H_2O$ and add five drops of $CuSO_4$ t.s. Record the color _____.

4. Check the USP/NF current edition for other specific color reactions that are the identifying characteristics on other sulfonamides such as sulfamerazine, sulfamethoxypyridazine, sulfisoxazole, and sulfamethazine.

## EXPERIMENT 67: NITRATE DETERMINATIONS OF SULFONAMIDES

In this experiment, you will perform assays of live sulfonamides (sulfadiazine, sulfacetamide, sulfamethazine, sulfamerazine, and sulfapyridine) using $NaNO_2$ as the titrant. Then calculate the equivalents by multiplying by the factor found in the table.

The procedure, to be repeated on each of the five drugs, is as follows: Weigh 500 mg of the drug and transfer to a beaker. Add 20 mL of 1 N HCl and 50 mL of $H_2O$. Stir until dissolved and cool to 15°C, by adding about 25 mg of crushed ice. Now slowly titrate with 0.1 M $NaNO_2$ solution using a magnetic stirrer until a blue color is produced. The weight (in mg) of the drug to which each milliliter of the 0.1 M $NaNO_2$ is equivalent is stated in the table below.

| Drug Name | mL of NaNO$_2$ Required | Equivalent Value | Total NO$_2$ |
|-----------|------------------------|------------------|--------------|
| Sulfadiazine | | 25.03 mg/mL | |
| Sulfacetamide | | 25.42 mg/mL | |
| Sulfamethazine | | 27.83 mg/mL | |
| Sulfamerazine | | 26.43 mg/mL | |
| Sulfapyridine | | 24.93 mg/mL | |

## EXPERIMENT 68: CHROMATOGRAPHIC PLATE IDENTIFICATION OF SULFONAMIDES

In this experiment, you will be identifying individual sulfonamides in a sulfa drug mixture by comparing $R_f$ values against standard preparations. Begin by preparing three solutions of standards by dissolving 100 mg of sulfadiazine in 50 mL of water using a small volumetric flask. Repeat this procedure with sulfathiazole and sulfamerazine, labeling each volumetric flask as 1, 2, or 3. Now prepare a fourth flask using 100 mg of the mixed drug (sample will be provided by your instructor) in 50 mL of water.

Then prepare three identical chromatographic plates already coated with silica gel mixture, by placing a 2:1 drop of each of the first three standard preparations about 2 cm apart along a line that is about 1.5 cm above the bottom of the plate. The fourth spot on the line of each plate will be a 2:1 drop of the test preparation taken from volumetric flask 4. The fifth and final spot along the line should consist of a 2:1 drop of each of the three standard preparations, thus making it a mixed standard.

While the spots are air-drying, the next step is to prepare three identical chromatographic chambers, each with a different solvent system. The first should consist of ethyl acetate, methanol, and a 25% aqueous solution of $NH_4OH$ in a 17:6:5 proportion. The second chamber should consist of hexane, chloroform, and butanol in a 1:1:1 mixture, and the third chamber of a solvent system, consisting of chloroform and methanol, in a 95:5 proportion. Place one prepared chromatographic plate in each chamber whose solvent level is just below the line of spots for each plate and develop the chromatogram until the solvent front has moved about three quarters the length of the plate.

Remove each plate from its development chamber, mark the solvent front, and allow the solvent to evaporate. Locate the spots by viewing the plates under short ultraviolet light. You may also spray the plates with a 1% solution of p-dimethylaminobenzaldehyde in dilute (5%) HCl and heat it at 110°C for 3 minutes or until bright yellow spots become visible.

The $R_f$ values of the yellow spots obtained from each test preparation should correspond to those obtained from the mixed standard preparations on the respective plates. The individual sulfonamides may be identified by comparisons in the $R_f$ values of the yellow spots obtained from test preparations and the individual standard preparations in the respective plates.

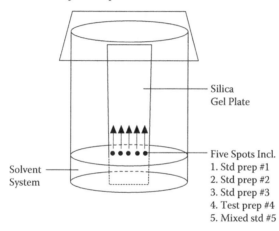

Record your findings in the table below.

|  | **$R_f$ Values** | | | | |
|  | **Standard Preparation 1** | **Standard Preparation 2** | **Standard Preparation 3** | **Test Preparation** | **Mixed Preparation** |
| Chamber #1 | | | | | |
| Chamber #2 | | | | | |
| Chamber #3 | | | | | |

## EXPERIMENT 69: ASSAY OF SULFADIAZINE TABLETS

Sulfadiazine finds use as an antibacterial agent in pharmacology and as an antimicrobial agent in veterinary practice. Its structure is

*Procedure for the assay*: Crush some sulfadiazine tablets using a mortar and pestle and place 0.5 g of the powder in a drying oven for 2 hours at 105° C. After this transfer, the powder to a 100 mL beaker, add 5 mL of 5.0 M HC1 plus 50 mL of water, and stir until the powder has dissolved. Cool to 15°C, add about 20 g of crushed ice and slowly titrate with 0.1 M $NaNO_2$, stirring vigorously until a blue color is produced immediately when a glass rod dipped into the titration solution is streaked on a smear of starch iodide t.s. When the titration is complete, the end point is reproducible after the mixture has been allowed to stand for 1 minute. Each milliliter of the 0.1 M $NaNO_2$ used is equivalent to 25.03 g of sulfadiazine ($C_{10}H_{10}N_4O_2S$).

Record results:

_____mL of 0.1 M $NaNO_2$ × 25.03 = _____ mg sulfadiazine

## SELECTED REFERENCES—SULFA DRUGS

Black, J. *Wilson and Gisvold's Organic Medicinal, and Pharmaceutical Chemistry*. Lippincott, 2011.

Brewster, R. Q. *Unitized Experiments in Organic Chemistry*. Van Nostrand, 1964.

Doerge, R. *Gisvold's Text of Organic, Medicinal, and Pharmaceutical Chemistry*. Lippincott, 1982.

Lemke, T. *Foye's Principles of Medicinal Chemistry*. Lippincott, 2010.

Lesch, J. *The First Miracle Drugs*. Oxford University Press, 2006.

Morrison, K. *Organic Chemistry*, Prentice Hall, 1991.

Patai, S. *The Chemistry of Sulphonic Acids*. Wiley, 1991.

Patrick, G. *An Introduction to Medicinal Chemistry*. Oxford University Press, 2009.

Vogel, A. J. *Practical Organic Chemistry*. Longmans, 1959.

# 20 Psychotropic Drugs

## PSYCHOTROPICS

The psychopharmacologic medications fall into two major categories as being either:

1. Tranquilizing agents
2. Psychomotor stimulants

Both classes find wide use in modern medical practice. Tranquilizing drugs are used in the treatment of neuroses and psychoses and include such drugs as chlorpromazine and perpanizine, which are both phenothiazine types, plus others such as probamate, adiol, and chlormethazanone, a methathiazanone derivative.

Psychomotor stimulants enhance alertness and result in an increased output of behavior and therefore are often used to treat patients in a depressive state. Drugs in this category include imipramine, amitriptyline, and doxepin plus the amphetamines.

Experiments in this section involve thin-layer chromatography on a psychotropic medication plus assays of two tricyclic antidepressants, amitriptyline and imipramine.

## EXPERIMENT 70: THIN-LAYER CHROMATOGRAPHY OF THIOTHIXENE

Psychotropics, drugs which alter the mind and behavior, have attracted the attention of man since the beginning of recorded history. They include such categories as tranquilizers, antipsychotics, psychomotor stimulants, and antidepressants. In this section, we will be performing assay investigations as well as chromatographic analysis on some representative samples.

*Procedure for Thiothixene TLC*: Dissolve 30 mg of thiothixene HCl (a substituted sulfonamide) in a 50:50 solution of 15 mL chloroform and 15 mL of methanol. Place on a magnetic stirrer for 10 minutes and then filter any solid that remains, using the clear supernatant for the test.

On a glass chromatographic plate coated with a 0.25 mm layer of silica gel, apply a drop of this solution using a micropipet. Allow the spot to dry and then develop a chromatogram using a solvent system consisting of a mixture of ethyl acetate, methanol, and diethyl amine (65:35:5 mL) until the solvent has moved about three quarters of the length of the plate.

Remove the plate from the developing chamber, mark the solvent front, and locate the spot on the plate by viewing under ultraviolet light. Then spray the plate lightly with acidified iodoplatinate spray.

Calculate the $R_f$ value for the thiothixene

$R_f$

## EXPERIMENT 71: ASSAY OF AMITRIPTYLINE HCL

Amitriptyline is a tricyclic antidepressant with a molecular structure as shown:

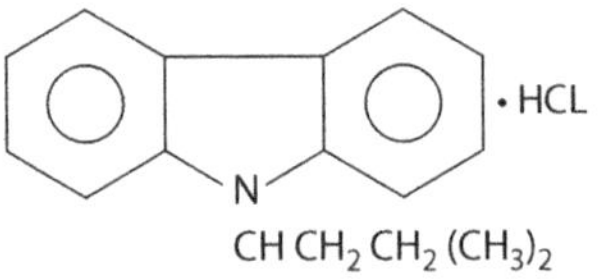

$C_{20} H_{23} N$ HCl
mol wt 313.87

•HCL

N
CH CH$_2$ CH$_2$ (CH$_3$)$_2$

*Procedure for Assay*: Dissolve 1 g of amitriptyline HCl in 30 mL of glacial acetic acid and warm slightly to effect solution. Cool, add 10 mL of mercuric acetate t.s., and titrate with 0.1 M perchloric acid ($HClO_4$) to a green end point.

Each milliliter of 0.1 M $HClO_4$ used is equivalent to 31.39 mg of amitriptyline HCl.

Record results:

_____ mg of $HClO_4$ × 31.39 _____ mg Amitr. HCl

Procedure for assay of imipramine HCL:
Imipramine HCl has the following structure:

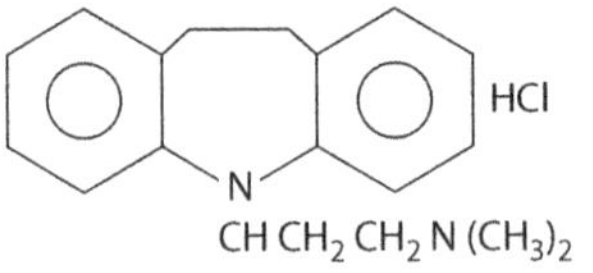

$C_{19} H_{25} N_2$ HCl
mol. wt 316.87

HCl

N
CH CH$_2$ CH$_2$ N (CH$_3$)$_2$

Dissolve 300 mg of imipramine HCl in 80 mL of glacial acetic acid and add 10 mL of mercuric acetate t.s. and 1 drop of crystal violet t.s. and titrate with 0.1 M $HClO_4$ to a blue end point. Each milliliter of $HClO_4$ equals 31.69 mg of imipramine HCl.

Record results:

_____ mL $HClO_4$ × 31.69 _____ mg Imipr. HCl

*Additional project*: List the various trade names and dosage strengths of amitriptyline and Imipramine. You may want to check with local pharmacies on this.

## SELECTED REFERENCES—PSYCHOTROPICS

Black, J. *Wilson & Gisvold's Organic, Medicinal, and Pharmaceutical Chemistry.* Lippincott, 2011.

Dubovsky, S. *Chemical Guide to Psychotropic Medicine.* Norton, 2005.

Gilbert, J. *Experimental Organic Chemistry.* Saunders, 1998.

Lemke, T. *Foye's Principle of Medicinal Chemistry.* Lippincott, 2010.

Patrick, G. *An Introduction to Medicinal Chemistry.* Oxford University Press, 2009.

Rogers, D. *Handbook of Essential Psychopharmacology.* American Psychiatric, 2005.

Thelheimer, W. *Synthetic Methods of Organic Chemistry.* Interscience, 1996.

Trost, B. *Comprehensive Organic Synthesis.* Pergamon, 1991.

# 21 Antibiotics

## ANTIBIOTICS

Antibiotics are natural products obtained from the growth of cultures of bacteria, molds, and soil actinomycetes. They are substances, which, in low concentrations, are destructive or inhibitory to microorganisms. The most famous of these is the drug penicillin produced by the mold growth, *Penicillium notatum*. All types of penicillin have the same basic structure but by substituting different radicals on the amino chain of the molecule, various types of penicillin may be produced.

Modern medicine also uses other antibiotics derived from eubacteriales such as bacitracin and gramicidin plus soil actinomycetes such as amphotericin and chloramphenicol. Additional antibiotics include erythromycin, neomycin, nystatin, and numerous tetracycline derivatives.

Experiments in this section include an investigation of color reactions of some selected antibiotics plus a demonstration of the antibacterial action of penicillin.

## EXPERIMENT 72: SOME COLOR REACTIONS OF ANTIBIOTICS

Antibiotics are natural products obtained from the growth of cultures of bacteria, molds, and soil actinomycetes. They may be defined as chemical substances derived from or produced by living organisms, which, in low concentrations, are destructive or inhibitory to microorganisms. In the beginning of antibiotic therapy, most antibiotics were obtained from natural processes, but now the majority are produced synthetically, thus paralleling the history of botanical drugs.

Erythromycin molecule

In this experiment, we will test the variant color reactions of three antibiotics (erythromycin, bacitracin, and neomycin sulfate) to common reagents and procedures.

To 5 mg of erythromycin add 2 mL of dilute $H_2SO_4$ and shake gently. Repeat this procedure using bacitracin and neomycin sulfate and record the color results.

---

### Results: Sulfuric Acid with Antibiotics

Erythromycin              _____
Bacitracin                _____
Neomycin sulfate          _____

---

To 3 mg of erythromycin, add 2 mL of acetone and 2 mL of dilute HCl. Repeat the procedure with bacitracin and neomycin sulfate and record results.

---

### Results: Acetone/HCl with Antibiotics

Erythromycin              _____
Bacitracin                _____
Neomycin sulfate          _____

---

To complete this experiment check, the microbiological and pharmaceutical literature and make a list some of the prescription drugs that contain any of these three antibiotics.

## EXPERIMENT 73: ANTIBACTERIAL ACTIVITY OF PENICILLIN G

The earliest antibiotic to be used commercially was penicillin from the *Penicillium notatum* mold, whose activity was reported by Sir Alexander Fleming in 1929. There are now many forms of penicillin employed in medicine today with slightly altered molecular structures.

In this experiment, you will be using potassium penicillin G to show its effect on bacterial cultures in broth.

Potassium Penicillin G

To each of four tubes containing 15 mL of fluid thioglycollate medium, add 1.0 mL of penicillin solution made to contain 100 USP units of potassium penicillin G in each milliliter. To each of two of the tubes containing the penicillin broth mixture, add sufficient penicillinase solution (1:10) to inactivate the penicillin. Incubate all four tubes at 37°C for 30 minutes, after which time inoculate them with 1.0 mL of a 1:1000 dilution of a culture of *Staphylococcus aureus* and incubate at 37°C for 24 hours. For a control, inoculate a tube of medium to which neither penicillin nor penicillinase has been added. Record results of tubes 1 through 5.

Tube 1. Broth with penicillin and penicillinase

_____

Tube 2. Broth with penicillin and penicillinase

_____

Tube 3. Broth with penicillin

_____

Tube 4. Broth with penicillin

_____

Tube 5. Broth with culture only

_____

Finally, check the pharmaceutical literature and make a list of some prescription drugs containing the various types of penicillin.

## SELECTED REFERENCES—ANTIBIOTICS

Abraham, E. P. *Biochemistry of Peptide and Steroid Antibiotics*. Wiley, 1957.
Bennett, P. *Clinical Pharmacology*. Elsevier, 2012.
Black, J. *Wilson and Gisvold's Organic, Medicinal, and Pharmaceutical Chemistry*. Lippincott, 2011.
Gilbert, D. *Sanford Guide to Antimicrobial Theory*. Antimicrobial, 2011.
Jukes, T. H. *Antibiotics in Nutrition*. Medical Encyclopedia, 1955.
Lemke, T. *Foye's Principles of Medicinal Chemistry*. Lippincott, 2010.
Patrick, G. *An Introduction to Medicinal Chemistry*. Lippincott, 2010.
Spoehr, H. A. *Fatty Acids: Antibacterials from Plants*. Carnegie, 1949.
Welch, H. A. *A Guide to Antibiotic Therapy*. Medical Encyclopedia, 1959.

# 22 Nucleic Acids

## NUCLEIC ACIDS

Nucleic acids are the molecules in cells that store and direct information for cellular growth and reproduction. The nucleic acid deoxyribonucleic cid (DNA), the genetic material in the nucleus of the cell, contains all the information needed for the development of a complete living system. Another type of nucleic acid, RNA, carries the genetic formation from the DNA to the ribosomes, which are cellular factories for the synthesis of protein.

In DNA and RNA, there are four nucleotides, each of which has three parts—a nitrogen base, a sugar, and a phosphate group. RNA consists of phosphoric acid, D-ribose, adenine, guanine, cytosine, and uracil. DNA is made up of phosphoric acid. D-2-deoxyribose, adenine, guanine, cytosine, and thymine.

The nucleic acids are slightly soluble in cold water, more readily soluble in hot water, and easily soluble in dilute alkali with the production of salts from which they can be separated by acids.

## EXPERIMENT 74: ISOLATION OF RNA

The experiments in this section include the isolation of RNA from baker's yeast and also the preparation of a cyclic nucleotide using sodium chloride. The method we will use in this experiment involves the direct extraction of yeast cells with phenol. The cells are not broken, and their walls permit the passage of RNA of low molecular weight into the extracting liquid.

*Procedure for extraction:* Mix 15 g of baker's yeast with 14 mL of phenol and 40 mL of water, and place on a magnetic stirrer for 1 hour. Then place the mixture in a 50 mL tube and centrifuge at 10,000 rpm for 12 minutes. About 25 mL of the supernatant can be collected by means of a rubber bulb attached to a suitable pipet. Then add 1 mL of phenol and 0.5 mL of water to the yeast extract with stirring and centrifugation as before. The upper phase should be separated as above. Then add 25 mL of 20% aqueous potassium acetate and 50 mL of 95% ethanol. Crude RNA is precipitated. Decant the supernatant and transfer the residue to a 15 mL tube by washing it from the large tubes with small amounts of 15% ethanol. After centrifugation, the ethanol should be decanted and the RNA dissolved in 3 mL of 0.1 HCl for column chromatography.

Prepare the column by mixing 1 g of DEAE-cellulose and stir with 40 mL of 0.1 HCl buffer. Pour the slurry into a 1 × 10 cm tube. Pour the liquid into the top of the column and add 40 mL of buffer and allow the liquid to percolate down the column. Finally, wash the column with 40 mL of buffer.

The RNA can then be precipitated from the eluate by adding 120 mL of 95% ethanol and separate the precipitate by centrifugation. Allow the s RNA to dry at room temperature.

Report results:

Weight of the soluble RNA _____ g.

Additional questions for investigation:

What part do polysaccharides play in RNA isolation?
Why is DEAE (diethylamino ethyl) cellulose used?

## EXPERIMENT 75: PREPARATION OF A CYCLIC NUCLEOTIDE

Ribonucleoside cyclic phosphates are the 2, 3 mono-phosphate esters of nucleosides. Two methods are generally used for the preparation of cyclic nucleotides, namely, the degradation of substances containing phosphodiester structures and the dehydration of nucleotides.

To 100 mg of uridine 2, 3 phosphate add 400 mg of DCC (dicyclohexylcarbodi-imide) in a 250 mL flask and add 8 mL of dimethylformamide. All the solid material should dissolve on gentle swirling. After a few minutes, the solution becomes cloudy and a precipitate appears. Add 40 mL of water to the reaction mixture, which should result in a pH of about 4. Filter, using a fritted glass filter, and wash the precipitate with 5 mL of water. Next, add 3 mL of 0.1 M NaOH and evaporate under reduced pressure until only a little diethyl formamide remains. Then add acetone to precipitate the 2, 3 phosphate and to extract the diethyl formamide and DCC.

Remote the acetone under reduced pressure and add 2 mL of water. Next, precipitate the barium salt by addition of 1 mL of barium acetate and a few drops of triachyl amine and 5 mL of ethanol.

Centrifuge the precipitate, wash with ethanol, followed by ether and dry in vacuo over phosphorus pentoxide.

Report results:

Weight of nucleotide precipitate _____ g.

## SELECTED REFERENCES—NUCLEIC ACIDS

Adams, R. L. *Biochemistry of the Nucleic Acids*. Chapman, 1992.
Blackburn, G. *Nucleic Acids*. RSC Publishing, 2002.
Cohn, W. *Progress in Nucleic Acid Research*. Academic, 1986.
Glick, D. *Methods of Biochemical Analysis*. Wiley, 1987.
Neidle, S. *Nucleic Acid Structure*. VCH Publications, 1987.
Revzin, A. *Footprinting of Nucleic Acid Structures*. Academic, 1991.
Tscheshe, H. *Modern Methods in Nucleic Acid Research*. Gruyter, 1990.
Ying, L. *Fundamental of Nucleic Acids*. Springer, 2009.
Zhang, L. H. *Medicinal Chemistry of Nucleic Acids*. Wiley, 2011.

# General Bibliography

Allport, N. *Chemistry and Pharmacy of Vegetable Drugs*. CPC, 1944.

Barton, D. *Comprehensive Natural Products Chemistry*. Pergamon, 1988.

Bean, H. *Advances in Pharmaceutical Sciences*. Academic, 1982.

Beckett, A. *Practical Pharmaceutical Chemistry*. Humanities, 1988.

Bennett, P. *Clinical Pharmacology*. Elsevier, 2012.

Briggs, T. *Biochemistry*. Springer-Verlag, 1991.

Clark, J. *Experimental Biochemistry*. Freeman, 1977.

Dickson, C. *Lab Experiments in Intro Chemistry*. ChemTec, 1996.

Dickson, C. *Medicinal Chemistry Laboratory Manual*. CRC, 1998.

Fiorkin, K. *Analytical Profiles of Drug Substances*. Academic, 1981.

Gleason, F. *Plant Biochemistry*. Jones & Bartlett, 2011.

Glick, D. *Methods of Biochemical Analysis*. Wiley, 1987.

Gokhale, S. B. *Pharmacognosy*. Pragmatic Books, 2008.

Gokhale, S. B. *Practical Pharmacognosy*. Prakashan, 2008.

Gurr, M. *Lipid Biochemistry*. Blackwell, 2002.

Halkerston, I. *Biochemistry*. Williams & Wilkins, 1990.

*Handbook of Chemistry and Physics*. 92nd ed. CRC, 2011.

Hegstad, L. *Essential Drug Design Calculations*. Appleton, 1989.

Heinrich, M. *Fundamentals of Pharmacognosy*. Elsevier, 2011.

Ikan, R. *Natural Products*. Academic, 1991.

Kapoor, L. D. *Handbook of Ayurvedic Medicine*. CRC, 2000.

Korikovasa, A. *Essentials of Medicinal Chemistry*. Wiley, 1988.

Kricka, L. *Analytical Methods*. Pergamon, 1988.

Ledrin, D. *Steroid Chemistry at a Glance*. Wiley, 2011.

Makowski, G. *Advances in Clinical Chemistry*. Elsevier, 2012.

Martin, A. *Physical Pharmacy*. Williams & Wilkins, 1995.

*The Merck Index*. 14th ed. Merck, 2011.

Nelson, D. *Principles of Biochemistry*. Freeman, 2008.

Patrick, G. *An Introduction to Medicinal Chemistry*. Oxford University Press, 2009.

*Physician's Desk Reference*. 65th ed. Medical Economics, 2013.

Ramstad, E. *Modern Pharmacognosy*. McGraw-Hill, 1957.

*Remington's Science and Practice of Pharmacy*. Lippincott, 2008.

Robbers, J. *Pharmacognosy Laboratory Guide*. Purdue University Press, 1981.

Roth, H. *Pharmaceutical Chemistry*. Prentice-Hall, 1991.

Stenesh, H. *Experimental Biochemistry*. Prentice-Hall, 1983.

Thelheimer, H. *Synthetic Methods of Organic Chemistry*. Wiley, 1999.

Tyler, V. *Pharmacognosy*. Lea & Febiger, 1988.

U.S. Pharmacopeia/National Formulary 35–30 USPC, 2011.

Wagner, H. *Economic and Medicinal Plant Research*. Academic, 1991.

Wigglesworth, J. *Biochemical Research*. Wiley, 1983.

Ying, L. *Fundamentals of Nucleic Acids*. Springer-Verlag, 2009.

# Index

Printed and bound by CPI Group (UK) Ltd, Croydon, CR0 4YY

23/10/2024

01777695-0006